PREPAREDNESS AND RESPONSE FOR A NUCLEAR OR RADIOLOGICAL EMERGENCY INVOLVING THE TRANSPORT OF RADIOACTIVE MATERIAL

The Agency's Statute was approved on 23 October 1956 by the Conference on the Statute of the IAEA held at United Nations Headquarters, New York; it entered into force on 29 July 1957. The Headquarters of the Agency are situated in Vienna. Its principal objective is "to accelerate and enlarge the contribution of atomic energy to peace, health and prosperity throughout the world".

IAEA SAFETY STANDARDS SERIES No. SSG-65

PREPAREDNESS AND RESPONSE FOR A NUCLEAR OR RADIOLOGICAL EMERGENCY INVOLVING THE TRANSPORT OF RADIOACTIVE MATERIAL

SPECIFIC SAFETY GUIDE

JOINTLY SPONSORED BY THE
INTERNATIONAL ATOMIC ENERGY AGENCY,
INTERNATIONAL CIVIL AVIATION ORGANIZATION,
INTERNATIONAL MARITIME ORGANIZATION

INTERNATIONAL ATOMIC ENERGY AGENCY
VIENNA, 2022

COPYRIGHT NOTICE

© IAEA, 2022

Printed by the IAEA in Austria
January 2022
STI/PUB/1960

IAEA Library Cataloguing in Publication Data

Names: International Atomic Energy Agency.
Title: Preparedness and response for a nuclear or radiological emergency involving the transport of radioactive material / International Atomic Energy Agency.
Description: Vienna : International Atomic Energy Agency, 2022. | Series: IAEA safety standards series, ISSN 1020–525X ; no. SSG-65 | Includes bibliographical references.
Identifiers: IAEAL 21-01450 | ISBN 978–92–0–127521–9 (paperback : alk. paper) | ISBN 978–92–0–127621–6 (pdf) | ISBN 978–92–0–127721–3 (epub)
Subjects: LCSH: Nuclear accidents — Management. | Ionizing radiation — Accidents. | Radioactive substances — Transportation. | Emergency management.
Classification: UDC 614.876 | STI/PUB/1960

FOREWORD

by Rafael Mariano Grossi
Director General

The IAEA's Statute authorizes it to "establish…standards of safety for protection of health and minimization of danger to life and property". These are standards that the IAEA must apply to its own operations, and that States can apply through their national regulations.

The IAEA started its safety standards programme in 1958 and there have been many developments since. As Director General, I am committed to ensuring that the IAEA maintains and improves upon this integrated, comprehensive and consistent set of up to date, user friendly and fit for purpose safety standards of high quality. Their proper application in the use of nuclear science and technology should offer a high level of protection for people and the environment across the world and provide the confidence necessary to allow for the ongoing use of nuclear technology for the benefit of all.

Safety is a national responsibility underpinned by a number of international conventions. The IAEA safety standards form a basis for these legal instruments and serve as a global reference to help parties meet their obligations. While safety standards are not legally binding on Member States, they are widely applied. They have become an indispensable reference point and a common denominator for the vast majority of Member States that have adopted these standards for use in national regulations to enhance safety in nuclear power generation, research reactors and fuel cycle facilities as well as in nuclear applications in medicine, industry, agriculture and research.

The IAEA safety standards are based on the practical experience of its Member States and produced through international consensus. The involvement of the members of the Safety Standards Committees, the Nuclear Security Guidance Committee and the Commission on Safety Standards is particularly important, and I am grateful to all those who contribute their knowledge and expertise to this endeavour.

The IAEA also uses these safety standards when it assists Member States through its review missions and advisory services. This helps Member States in the application of the standards and enables valuable experience and insight to be shared. Feedback from these missions and services, and lessons identified from events and experience in the use and application of the safety standards, are taken into account during their periodic revision.

I believe the IAEA safety standards and their application make an invaluable contribution to ensuring a high level of safety in the use of nuclear technology. I encourage all Member States to promote and apply these standards, and to work with the IAEA to uphold their quality now and in the future.

PREFACE

In March 2015, the IAEA's Board of Governors approved a Safety Requirements publication, IAEA Safety Standards Series No. GSR Part 7, Preparedness and Response for a Nuclear or Radiological Emergency, which was jointly sponsored by 13 international organizations. GSR Part 7 establishes requirements for an adequate level of preparedness for and response to a nuclear or radiological emergency, irrespective of the initiator of the emergency.

The Convention on Early Notification of a Nuclear Accident and the Convention on Assistance in the Case of a Nuclear Accident or Radiological Emergency ('the Assistance Convention'), adopted in 1986, place specific obligations on the States Parties and on the IAEA. Under Article 5a(ii) of the Assistance Convention, one function of the IAEA is to "collect and disseminate to States Parties and Member States information concerning:…methodologies, techniques and available results of research relating to response to nuclear accidents or radiological emergencies".

In March 2018, the IAEA's Board of Governors approved the most recent edition of the IAEA's Transport Regulations, which were issued as IAEA Safety Standards Series No. SSR-6 (Rev. 1), Regulations for the Safe Transport of Radioactive Material, 2018 Edition. The IAEA's Transport Regulations establish requirements that must be satisfied to ensure safety and to protect people, property and the environment from harmful effects of ionizing radiation during the transport of radioactive material.

The IAEA General Conference, in resolution GC(59)/RES/9 requested "the Secretariat, Member States and relevant international organizations to emphasize the specific challenges and requirements for efficient international cooperation in response to nuclear and radiological incidents and emergencies relating to the transport of radioactive material".

This Safety Guide is intended to assist Member States in the application of GSR Part 7 and of the Transport Regulations. It provides guidance and recommendations on emergency arrangements for the transport of radioactive material. The recommendations in this Safety Guide are aimed at States, regulatory bodies and response organizations, including consignors, carriers and consignees.

This Safety Guide supersedes IAEA Safety Standards Series No. TS-G-1.2 (ST-3), Planning and Preparing for Emergency Response to Transport Accidents Involving Radioactive Material.

The IAEA, the International Civil Aviation Organization and the International Maritime Organization are joint sponsors of this Safety Guide.

THE IAEA SAFETY STANDARDS

BACKGROUND

Radioactivity is a natural phenomenon and natural sources of radiation are features of the environment. Radiation and radioactive substances have many beneficial applications, ranging from power generation to uses in medicine, industry and agriculture. The radiation risks to workers and the public and to the environment that may arise from these applications have to be assessed and, if necessary, controlled.

Activities such as the medical uses of radiation, the operation of nuclear installations, the production, transport and use of radioactive material, and the management of radioactive waste must therefore be subject to standards of safety.

Regulating safety is a national responsibility. However, radiation risks may transcend national borders, and international cooperation serves to promote and enhance safety globally by exchanging experience and by improving capabilities to control hazards, to prevent accidents, to respond to emergencies and to mitigate any harmful consequences.

States have an obligation of diligence and duty of care, and are expected to fulfil their national and international undertakings and obligations.

International safety standards provide support for States in meeting their obligations under general principles of international law, such as those relating to environmental protection. International safety standards also promote and assure confidence in safety and facilitate international commerce and trade.

A global nuclear safety regime is in place and is being continuously improved. IAEA safety standards, which support the implementation of binding international instruments and national safety infrastructures, are a cornerstone of this global regime. The IAEA safety standards constitute a useful tool for contracting parties to assess their performance under these international conventions.

THE IAEA SAFETY STANDARDS

The status of the IAEA safety standards derives from the IAEA's Statute, which authorizes the IAEA to establish or adopt, in consultation and, where appropriate, in collaboration with the competent organs of the United Nations and with the specialized agencies concerned, standards of safety for protection of health and minimization of danger to life and property, and to provide for their application.

With a view to ensuring the protection of people and the environment from harmful effects of ionizing radiation, the IAEA safety standards establish fundamental safety principles, requirements and measures to control the radiation exposure of people and the release of radioactive material to the environment, to restrict the likelihood of events that might lead to a loss of control over a nuclear reactor core, nuclear chain reaction, radioactive source or any other source of radiation, and to mitigate the consequences of such events if they were to occur. The standards apply to facilities and activities that give rise to radiation risks, including nuclear installations, the use of radiation and radioactive sources, the transport of radioactive material and the management of radioactive waste.

Safety measures and security measures[1] have in common the aim of protecting human life and health and the environment. Safety measures and security measures must be designed and implemented in an integrated manner so that security measures do not compromise safety and safety measures do not compromise security.

The IAEA safety standards reflect an international consensus on what constitutes a high level of safety for protecting people and the environment from harmful effects of ionizing radiation. They are issued in the IAEA Safety Standards Series, which has three categories (see Fig. 1).

Safety Fundamentals

Safety Fundamentals present the fundamental safety objective and principles of protection and safety, and provide the basis for the safety requirements.

Safety Requirements

An integrated and consistent set of Safety Requirements establishes the requirements that must be met to ensure the protection of people and the environment, both now and in the future. The requirements are governed by the objective and principles of the Safety Fundamentals. If the requirements are not met, measures must be taken to reach or restore the required level of safety. The format and style of the requirements facilitate their use for the establishment, in a harmonized manner, of a national regulatory framework. Requirements, including numbered 'overarching' requirements, are expressed as 'shall' statements. Many requirements are not addressed to a specific party, the implication being that the appropriate parties are responsible for fulfilling them.

Safety Guides

Safety Guides provide recommendations and guidance on how to comply with the safety requirements, indicating an international consensus that it

---

[1] See also publications issued in the IAEA Nuclear Security Series.

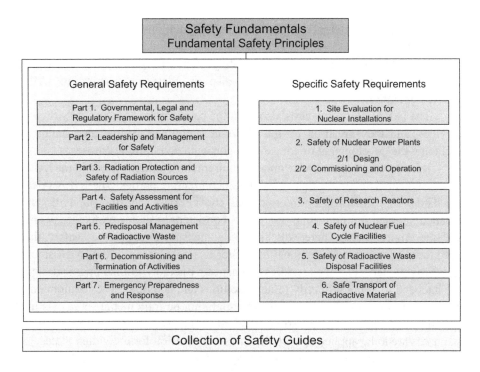

FIG. 1. The long term structure of the IAEA Safety Standards Series.

is necessary to take the measures recommended (or equivalent alternative measures). The Safety Guides present international good practices, and increasingly they reflect best practices, to help users striving to achieve high levels of safety. The recommendations provided in Safety Guides are expressed as 'should' statements.

APPLICATION OF THE IAEA SAFETY STANDARDS

The principal users of safety standards in IAEA Member States are regulatory bodies and other relevant national authorities. The IAEA safety standards are also used by co-sponsoring organizations and by many organizations that design, construct and operate nuclear facilities, as well as organizations involved in the use of radiation and radioactive sources.

The IAEA safety standards are applicable, as relevant, throughout the entire lifetime of all facilities and activities — existing and new — utilized for peaceful purposes and to protective actions to reduce existing radiation risks. They can be used by States as a reference for their national regulations in respect of facilities and activities.

The IAEA's Statute makes the safety standards binding on the IAEA in relation to its own operations and also on States in relation to IAEA assisted operations.

The IAEA safety standards also form the basis for the IAEA's safety review services, and they are used by the IAEA in support of competence building, including the development of educational curricula and training courses.

International conventions contain requirements similar to those in the IAEA safety standards and make them binding on contracting parties. The IAEA safety standards, supplemented by international conventions, industry standards and detailed national requirements, establish a consistent basis for protecting people and the environment. There will also be some special aspects of safety that need to be assessed at the national level. For example, many of the IAEA safety standards, in particular those addressing aspects of safety in planning or design, are intended to apply primarily to new facilities and activities. The requirements established in the IAEA safety standards might not be fully met at some existing facilities that were built to earlier standards. The way in which IAEA safety standards are to be applied to such facilities is a decision for individual States.

The scientific considerations underlying the IAEA safety standards provide an objective basis for decisions concerning safety; however, decision makers must also make informed judgements and must determine how best to balance the benefits of an action or an activity against the associated radiation risks and any other detrimental impacts to which it gives rise.

DEVELOPMENT PROCESS FOR THE IAEA SAFETY STANDARDS

The preparation and review of the safety standards involves the IAEA Secretariat and five Safety Standards Committees, for emergency preparedness and response (EPReSC) (as of 2016), nuclear safety (NUSSC), radiation safety (RASSC), the safety of radioactive waste (WASSC) and the safe transport of radioactive material (TRANSSC), and a Commission on Safety Standards (CSS) which oversees the IAEA safety standards programme (see Fig. 2).

All IAEA Member States may nominate experts for the Safety Standards Committees and may provide comments on draft standards. The membership of the Commission on Safety Standards is appointed by the Director General and includes senior governmental officials having responsibility for establishing national standards.

A management system has been established for the processes of planning, developing, reviewing, revising and establishing the IAEA safety standards.

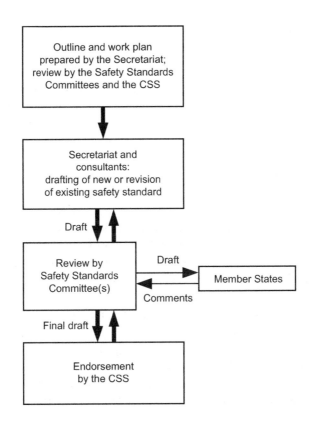

FIG. 2. The process for developing a new safety standard or revising an existing standard.

It articulates the mandate of the IAEA, the vision for the future application of the safety standards, policies and strategies, and corresponding functions and responsibilities.

INTERACTION WITH OTHER INTERNATIONAL ORGANIZATIONS

The findings of the United Nations Scientific Committee on the Effects of Atomic Radiation (UNSCEAR) and the recommendations of international expert bodies, notably the International Commission on Radiological Protection (ICRP), are taken into account in developing the IAEA safety standards. Some safety standards are developed in cooperation with other bodies in the United Nations system or other specialized agencies, including the Food and Agriculture Organization of the United Nations, the United Nations Environment Programme, the International Labour Organization, the OECD Nuclear Energy Agency, the Pan American Health Organization and the World Health Organization.

INTERPRETATION OF THE TEXT

Safety related terms are to be understood as defined in the IAEA Safety Glossary (see https://www.iaea.org/resources/safety-standards/safety-glossary). Otherwise, words are used with the spellings and meanings assigned to them in the latest edition of The Concise Oxford Dictionary. For Safety Guides, the English version of the text is the authoritative version.

The background and context of each standard in the IAEA Safety Standards Series and its objective, scope and structure are explained in Section 1, Introduction, of each publication.

Material for which there is no appropriate place in the body text (e.g. material that is subsidiary to or separate from the body text, is included in support of statements in the body text, or describes methods of calculation, procedures or limits and conditions) may be presented in appendices or annexes.

An appendix, if included, is considered to form an integral part of the safety standard. Material in an appendix has the same status as the body text, and the IAEA assumes authorship of it. Annexes and footnotes to the main text, if included, are used to provide practical examples or additional information or explanation. Annexes and footnotes are not integral parts of the main text. Annex material published by the IAEA is not necessarily issued under its authorship; material under other authorship may be presented in annexes to the safety standards. Extraneous material presented in annexes is excerpted and adapted as necessary to be generally useful.

CONTENTS

1. INTRODUCTION

BACKGROUND

1.1. Under Article 5(a)(ii) of the Convention on Assistance in the Case of a Nuclear Accident or Radiological Emergency [1], one function of the IAEA is to "collect and disseminate to States Parties and Member States information concerning:…methodologies, techniques and available results of research relating to response to nuclear accidents or radiological emergencies".

1.2. IAEA Safety Standards Series No. GSR Part 7, Preparedness and Response for a Nuclear or Radiological Emergency [2], establishes requirements for an adequate level of preparedness and response for a nuclear or radiological emergency[1], irrespective of the initiator of the emergency.

1.3. IAEA Safety Standards Series No. SSR-6 (Rev. 1), Regulations for the Safe Transport of Radioactive Material, 2018 Edition [3] (hereinafter referred to as the 'Transport Regulations'), establishes requirements to be complied with by competent authorities, package designers, consignors, carriers and consignees. Meeting these requirements ensures a high level of safety for the transport of radioactive material. However, events during transport can occur, and some of these events might lead to a nuclear or radiological emergency. Advance planning and preparation are necessary to provide an efficient and effective response to such emergencies. Consequently, the Transport Regulations [3] require arrangements for emergency preparedness and response for the transport of radioactive material.

1.4. Packages used for the transport of radioactive material are designed with a graded approach to meet requirements that include consideration of the effects on the package of prescribed accident conditions of transport. Consequently, most emergencies during transport have limited radiological consequences and can be resolved in a relatively short period. The emergency response may last only hours or days. However, this Safety Guide considers a wide range of possible emergencies, including those associated with very low probability events that might have significant radiological consequences.

---

[1] A nuclear or radiological emergency is an emergency in which there is, or is perceived to be, a hazard due to (i) the energy resulting from a nuclear chain reaction or from the decay of the products of a chain reaction or (ii) radiation exposure [2].

1.5. This Safety Guide supersedes IAEA Safety Standards Series No. TS-G-1.2 (ST-3), Planning and Preparing for Emergency Response to Transport Accidents Involving Radioactive Material[2].

1.6. Unless otherwise specified, terms used in this Safety Guide are as defined in the IAEA Safety Glossary [4]. For reasons of brevity, in this Safety Guide the term 'emergency' is intended to mean a nuclear or radiological emergency, unless otherwise specified.

OBJECTIVE

1.7. The objective of this publication is to provide recommendations on emergency preparedness and response for the transport of radioactive material. These recommendations form the basis of achieving the goals of emergency response described in GSR Part 7 [2].

1.8. The recommendations in this Safety Guide are aimed at States[3], regulatory bodies and response organizations, including consignors, carriers and consignees.

1.9. This Safety Guide should be used in conjunction with the requirements established in GSR Part 7 [2] and the Transport Regulations [3] for emergency preparedness and response for the transport of radioactive material[4], with due account taken of the recommendations provided in IAEA Safety Standards Series Nos GS-G-2.1, Arrangements for Preparedness for a Nuclear or Radiological Emergency [5]; GSG-2, Criteria for Use in Preparedness and Response for a Nuclear or Radiological Emergency [6]; and GSG-11, Arrangements for the Termination of a Nuclear or Radiological Emergency [7].

---

[2] INTERNATIONAL ATOMIC ENERGY AGENCY, Planning and Preparing for Emergency Response to Transport Accidents Involving Radioactive Material, IAEA Safety Standards Series No. TS-G-1.2 (ST-3), IAEA, Vienna (2002).

[3] GSR Part 7 [2] uses the term 'State', while the Transport Regulations [3] use the term 'country'. In this Safety Guide, the terms 'State' and 'country' are used synonymously.

[4] In this Safety Guide, the term 'radioactive material' means "Material designated in national law or by a *regulatory body* as being subject to *regulatory control* because of its *radioactivity*" [4]. This matches the definition in GSR Part 7 [2] and includes all material that falls within the definition of radioactive material in the Transport Regulations [3]. The term 'radioactive material' also includes nuclear material, as defined in the IAEA Nuclear Security Series.

SCOPE

1.10. This Safety Guide considers emergency preparedness and response for the transport of radioactive material, irrespective of the initiator of the emergency, which could be a natural event, a human error, a mechanical or other failure, or a nuclear security event [2].

1.11. The scope of this Safety Guide is limited to transport activities under emergency preparedness category IV[5], as defined in table 1 of GSR Part 7 [2].

1.12. This Safety Guide does not apply to events occurring during the transport of radioactive material that do not initiate a nuclear or radiological emergency, for example a vehicle involved in a minor traffic accident or an accident involving radioactive material categorized as LSA-I, as defined in para. 409(a), or SCO-I, as defined in para. 413(a) of the Transport Regulations [3]. An accident during the transport of these types of material, whether packaged or unpackaged, is unlikely to lead to a nuclear or radiological emergency.

1.13. This Safety Guide does not apply to emergencies involving the movement of radioactive material fully within the boundaries of authorized facilities. Such emergencies should be addressed as part of the on-site emergency arrangements for the facility, consistent with the relevant requirements in GSR Part 7 [2].

---

[5] Emergency preparedness category IV is defined as follows (footnotes omitted):

"Activities and acts that could give rise to a nuclear or radiological emergency that could warrant protective actions and other response actions to achieve the goals of emergency response in accordance with international standards in an unforeseen location. These activities and acts include: (a) transport of nuclear or radioactive material and other authorized activities involving mobile dangerous sources such as industrial radiography sources, nuclear powered satellites or radioisotope thermoelectric generators; and (b) theft of a dangerous source and use of a radiological dispersal device or radiological exposure device. This category also includes: (i) detection of elevated radiation levels of unknown origin or of commodities with contamination; (ii) identification of clinical symptoms due to exposure to radiation; and (iii) a transnational emergency that is not in category V arising from a nuclear or radiological emergency in another State. Category IV represents a level of hazard that applies for all States and jurisdictions" [2].

1.14. This Safety Guide does not address measures specific to nuclear security; such measures are addressed in the publications of the IAEA Nuclear Security Series. The interface with nuclear security response measures is addressed in Section 5.

1.15. Emergency preparedness and response for transport should consider all hazards that might be present. The hazards may include radiological hazards, other hazards from the shipment and operational hazards at the emergency site. Non-radiological hazards are outside the scope of this Safety Guide and are considered only when they might affect the response to radiological hazards.

STRUCTURE

1.16. Section 2 describes the overall national emergency arrangements and framework for emergency preparedness and response for transport. It defines the roles and responsibilities of States, regulatory bodies, consignors, carriers and radiological assessors. Section 3 describes preparedness and response elements, including the preparedness stage, the concept of operations, and training, drills and exercises. Section 4 describes specific considerations for each mode of transport, which can be considered in the overall context of the concept of operations described in Section 3. Section 5 describes the interface with nuclear security and provides references to relevant publications in the IAEA Nuclear Security Series.

1.17. Appendix I sets out considerations for States that are developing a national capability for emergency preparedness and response for transport. Appendix II describes the types of initiating event that could lead to an emergency. Annex I provides background information on the requirements of the Transport Regulations [3] that are relevant to emergency response. Annex II provides an example event notification form. Annex III provides a template for the emergency response plan for carriers or consignors. Annex IV describes possible radiological consequences of postulated emergency situations.

2. NATIONAL ARRANGEMENTS AND FRAMEWORK

2.1. This section provides recommendations on establishing and maintaining arrangements for emergency preparedness and response for the transport of radioactive material. Additional guidance on developing arrangements is contained in Appendix I.

2.2.	The arrangements described in this section are intended to help States achieve the goals of emergency preparedness and response, as defined in para. 3.2 of GSR Part 7 [2], and are intended to be part of an effective governmental, legal and regulatory framework (see IAEA Safety Standards Series No. GSR Part 1 (Rev. 1), Governmental, Legal and Regulatory Framework for Safety [8]) in relation to emergency preparedness and response for transport.

EMERGENCY MANAGEMENT SYSTEM

2.3.	In accordance with Requirement 1 of GSR Part 7 [2], the government is required to establish and maintain an emergency management system. This system should integrate all relevant elements (e.g. organizational structure, resources, policies, processes) into one coherent system to enable the organizations involved to set clear goals and strategies in emergency preparedness and response for any emergency during transport in an effective manner, commensurate with the results of the hazard assessment.

2.4.	An incident during the transport of radioactive material might occur in any location. This level of hazard applies to all States and jurisdictions, as defined in emergency preparedness category IV in table 1 of GSR Part 7 [2].

2.5.	In accordance with para. 4.10 of GSR Part 7 [2], the government is required to establish a national coordinating mechanism designed "To ensure that roles and responsibilities are clearly specified and are understood by operating organizations, response organizations and the regulatory body". This mechanism should also ensure that emergency arrangements are established in accordance with a graded approach.

2.6.	The national coordinating mechanism should identify all the response organizations and agencies involved in emergency preparedness and response for the transport of radioactive material at the national and local levels, including any foreign consignors, carriers and consignees.

2.7.	The national coordinating mechanism should include the relevant competent authorities for emergency preparedness and response, for transport safety and for transport security, in accordance with the relevant national circumstances. These competent authorities might be the same organization.

2.8.	Within the national coordinating mechanism, various organizations may have distinct responsibilities concerning a transport emergency. Where practicable, one organization should be assigned responsibility for each aspect of emergency preparedness and response.

2.9. Emergency arrangements should be put in place for a transport emergency involving consignors or carriers, including foreign consignors and carriers operating within or through the State. The arrangements of foreign consignors and carriers should comply with national regulations and should be compatible with the actions of national response organizations, including coordination and communication.

2.10. In accordance with para. 5.7 of GSR Part 7 [2], at the preparedness stage, arrangements are required to be made for the establishment of a unified command and control system with clear authority and responsibility[6] to direct the response at the site area[7] during an emergency, including the response of public and private response organizations that may be present at a transport emergency.

ROLES AND RESPONSIBILITIES

2.11. In accordance with Requirement 2 of GSR Part 7 [2], at the preparedness stage, the roles and responsibilities of all organizations involved in emergency preparedness and response for transport — the government, the response organizations (national and local), and emergency workers, including first responders, radiological assessors, carriers and consignors — are required to be clearly specified and clearly assigned. In some cases, consignees may also have responsibilities in the event of a transport emergency.

2.12. Since the preparedness and response actions for a transport emergency involving any class of dangerous goods have much in common, the aspects that are specific to nuclear or radiological emergencies should be developed and incorporated into the overall emergency management system under the all-hazards approach: see para. 4.3 of GSR Part 7 [2].

---

[6] This authority and responsibility would typically be assigned to an individual in the organization that has the primary role during each phase of the response. The authority and responsibility may be transferred between organizations as the emergency response progresses.

[7] For a transport emergency, the term 'site area' refers to the inner cordoned off area established by first responders around a suspected hazard [2].

Government

2.13. Paragraph 4.5 of GSR Part 7 [2] states:

"The government shall make adequate preparations to anticipate, prepare for, respond to and recover from a nuclear or radiological emergency at the operating organization, local, regional and national levels, and also, as appropriate, at the international level."

With regard to the transport of radioactive material, the operating organization includes consignors, carriers and consignees, as appropriate.

2.14. The relevant governmental bodies should ensure that:

(a) Specific provisions regarding emergency preparedness and response for the transport of radioactive material are taken into account by the national coordinating mechanism (see para. 4.10 of GSR Part 7 [2]), which should include representatives of the transport safety competent authority. These provisions should be kept up to date.

(b) The national regulatory requirements for emergency preparedness and response for consignors and carriers, including foreign consignors and carriers operating within or through the State, are defined and included in the international regulatory framework for dangerous goods, as appropriate.

(c) Arrangements are in place to respond to the loss or theft of radioactive material during transport. If radioactive material has been lost during transport, it should be treated as material out of regulatory control. Recommendations and guidance on nuclear and other radioactive material out of regulatory control are provided in the IAEA Nuclear Security Series.

2.15. In developing emergency arrangements, the relevant government bodies, including the competent authorities for transport safety, transport security and emergency preparedness and response, should do the following:

(a) Ensure that legislation defines the areas of responsibility and the functions of the various national authorities that have expertise in transport safety, transport security and emergency preparedness and response;

(b) Ensure that the national coordinating mechanism includes the transport safety competent authority;

(c) Define the responsibilities of national and local governments for a transport emergency, which could occur in any location;

(d) Identify consignors, carriers and consignees involved in the regular transport of consignments of radioactive material within or through the State, to enable an accurate hazard assessment to be performed;

(e) Ensure that any persons with relevant technical expertise (e.g. experts on the package design, radiological assessors) are available in the event of a transport emergency;

(f) Identify the authorities and organizations to be notified when an incident occurs during the transport of radioactive material, and establish notification procedures;

(g) Ensure the periodic review, testing and updating of response organization plans, which may include those of private response organizations;

(h) Establish proper training, drill and exercise programmes that include all response organizations;

(i) Consider establishing arrangements with the governments in relevant States, including neighbouring States, for a transport emergency that extends beyond national boundaries;

(j) Specify the responsibility for the provision and coordination of public information in the event of a transport emergency, including the role of the consignor and carrier;

(k) Ensure that the necessary human, financial and other resources are available to prepare for and deal with a transport emergency;

(l) Ensure that arrangements are in place for the compensation of victims for damage due to a transport emergency.

2.16. Local governments should develop emergency arrangements for a transport emergency based on the national requirements and the national hazard assessment. These arrangements should address the ability to recognize a consignment of radioactive material, provide familiarity with basic safety precautions, and specify whom to notify. These arrangements should include the deployment and operation of the local government's own resources.

Consignors and carriers

2.17. The consignor has the primary responsibility for ensuring that adequate emergency arrangements are in place for a given shipment of radioactive material and that those arrangements follow the national emergency arrangements of all the States relevant to the shipment. Some aspects of this responsibility may be assigned to the carrier. The States relevant to the shipment include the following, as appropriate:

(a) The flag State of the conveyance;

(b) The State of the consignor;

(c)   The State of the consignee;

(d)   States with land, air or territorial waters through which the shipment transits.

2.18. The consignor should ensure that, before carriers undertake the transport of a consignment of radioactive material, they are provided with the instructions to be followed in the event of a transport emergency.

2.19. The consignor should provide instructions, when appropriate, for any specific environmental conditions relevant to the shipment and its route (e.g. remote locations or areas with poor mobile phone reception, road tunnels or risk of severe adverse weather conditions).

2.20. The consignor should confirm that the carrier has made emergency arrangements with relevant organizations, which could include private companies, throughout the duration of the shipment, through all States and jurisdictions, taking account of the possibility of multiple modes of transport. These arrangements should be applied in a graded approach, considering the consignment and based on aspects such as distances, languages or applicable jurisdictional requirements for the shipment.

2.21. The carrier should ensure that written emergency instructions applicable to the consignment of radioactive material being transported are carried on board the conveyance. In addition, efforts should be made by the carrier to ensure that this emergency information will be available to the first responders, even if the carrier personnel are incapacitated.

2.22. During an emergency, the consignor or the carrier may need to communicate with the media and the public, depending on the national arrangements in place. When required, the information should be shared between the different authorities and response organizations involved, before being published, to help ensure that accurate and consistent information is provided [9].

Radiological assessor

2.23. In some cases, emergency services for conventional emergencies are sufficient to respond to a nuclear or radiological emergency. However, if it is suspected that the integrity of a package might have been compromised, a radiological assessor with specialized expertise may be needed to respond to the emergency. The emergency arrangements should include provisions for identifying the necessary expertise and skills of radiological assessors[8] and for the timely involvement of appropriate radiological assessors in the response.

2.24. The role of the radiological assessor, which can be fulfilled by either an individual or a team, is to perform radiological surveys, dose assessments and contamination control; to ensure the radiation protection of emergency workers; and to advise on protective actions and other response actions. This role may be fulfilled remotely or at the emergency site, depending on the emergency situation.

2.25. Radiological assessors should be trained and qualified in their necessary functions, including assessing radiation safety, assessing package containment, making dose rate and contamination measurements, and advising on protective actions. Depending on the results of the hazard assessment, the radiological assessor may also need to be trained in the assessment and prevention of criticality.

2.26. Capabilities to communicate with the radiological assessors should be continuously available so that they can be notified of an emergency in which their expertise is needed.

2.27. If the emergency arrangements rely on the presence of radiological assessors at the site of the emergency, such assessors should be able to reach the site area within an appropriate response time. This could be achieved by identifying assessors and equipment across the State or jurisdiction or by having a preidentified means of transport for a centralized team of assessors and equipment, to ensure their timely attendance at the site of the emergency.

2.28. Radiological assessors should be prepared and equipped to undertake the following:

(a) Travel to the site, if needed, with the appropriate equipment, within the time specified in the emergency arrangements;
(b) Integrate into the unified command and control system and coordinate with response organizations;
(c) Operate in emergency conditions, if necessary, while ensuring their own protection from radiological and non-radiological hazards at all times;
(d) Evaluate the radiological hazard resulting from the radioactive material, through measurements, observations, sampling and other methods, as appropriate;

---

[8] Depending on the situation and the national arrangements, the radiological assessor may come from government, technical support organizations, or consignors and carriers.

(e)   Advise on the appropriate steps to minimize the radiation exposure of people;

(f)   Minimize the spread of radioactive contamination;

(g)   Assess the package safety functions and provide a prognosis of their future development;

(h)   Provide technical information and advice to the appropriate authorities and response organizations to help in the emergency response.

HAZARD ASSESSMENT

2.29. Requirement 4 of GSR Part 7 [2] states that: "**The government shall ensure that a hazard assessment is performed to provide a basis for a graded approach in preparedness and response for a nuclear or radiological emergency.**" The potential hazards associated with a transport emergency are required to be identified to provide a basis for establishing emergency arrangements commensurate with the potential consequences of an incident. The hazard assessment is required to identify postulated initiating events and assess the potential consequences for people, property and the environment. Possible initiating events are listed in Appendix II. Possible consequences of an emergency are listed in Annex IV. The hazard assessment provides the basis for a graded approach and allows the development of emergency arrangements commensurate with the potential consequences.

2.30. The hazard assessment should be based on information provided by consignors, carriers, local governments and competent authorities.

2.31. In accordance with paras 4.18 and 4.24 of GSR Part 7 [2], the potential consequences of the identified hazards — radiation hazards, as well as non-radiation-related hazards that might impair the emergency response — are required to be assessed. This assessment should include an evaluation of the potential external dose, the potential for intakes of radioactive material and an evaluation of the associated internal doses that could be received.

2.32. There are multiple values associated with the activity of radioactive material that are used for different purposes in the transport of radioactive material and in emergency preparedness and response. The A_1 and A_2 values[9] defined in the Transport Regulations [3] are used "to determine the activity limits for the requirements of these Regulations." In emergency preparedness and response, D values have been developed for individual radionuclides to specify the activity of a source that, if not under control, could cause severe deterministic effects in a range of scenarios that include both external exposure from an unshielded source and internal exposure following dispersal of the source material [10]. Thus, the A_1 and A_2 values are used to determine required package types, with the goal of applying the graded approach of the Transport Regulations [3] to shipments and withstanding accident conditions of transport. In contrast, D values should be used in determining the extent of the emergency arrangements — using a graded approach consistent with the recommendations provided in GS-G-2.1 [5] — that are necessary to avoid or minimize severe deterministic effects.

2.33. The different types of package and their radioactive contents transported within or through the State should be considered in the hazard assessment. The hazard assessment should consider events leading to single or combined failures of the package safety functions (e.g. containment, protection against external radiation, prevention of damage caused by heat, prevention of criticality), the risks arising from the transport mode and route, and the risk of human error.

2.34. For the assessment of identified hazards, other external conditions that could hinder or impair the response capability should be taken into account when their combination with a transport emergency is foreseeable. Such external conditions include the following:

(a)   A conventional emergency such as an earthquake, hurricane, flood or severe weather at sea (see para. 4.20(b) of GSR Part 7 [2]);
(b)   Another simultaneous emergency affecting a nearby facility;
(c)   Non-radiological hazards arising during the transport emergency.

---

[9]   The Transport Regulations [3] define the A_1 and A_2 values as follows:

"A_1 shall mean the activity value of *special form radioactive material* that is listed in Table 2 or derived in Section IV [of the Transport Regulations] and is used to determine the activity limits for the requirements of these Regulations. A_2 shall mean the activity value of *radioactive material*, other than *special form radioactive material*, that is listed in Table 2 or derived in Section IV [of the Transport Regulations] and is used to determine the activity limits for the requirements of these Regulations."

2.35. The graded approach to the requirements for package classification and design, as described in the Transport Regulations [3], has been developed in part to limit the radiation exposure of workers. However, foreseeable events, even those with low probabilities, in which the package can be compromised beyond the design requirements should be considered in the hazard assessment. Events of this type include the following:

(a) Operational errors associated with human and organizational factors in package preparation, resulting in excessive dose rates. Examples include making an error in preparation for shipment, forgetting to engage a closure bolt or failing to install the required shielding.

(b) Exceptional environmental loadings, such as tunnel fires, burying in soft soil, covering with debris, high energy crushing (exceeding the energy of the 9 m drop test), sharp impacts (e.g. impact from a forklift truck) and airplane crashes (except for Type C packages).

2.36. Modalities of transport should also be considered when identifying initiating events and potential consequences. Factors for consideration include route, nearby infrastructure, terrain, distance, timing, seasonal weather and sensitive environments (e.g. environments that contain local food and water supplies). Additional factors, including the frequency of transport, should be taken into account when applying a graded approach to identifying the planning basis for emergency preparedness and response.

2.37. In accordance with para. 4.25 of GSR Part 7 [2], a periodic review of the hazard assessment is required to be undertaken to ensure that any major change in transport activities is adequately considered and that existing arrangements remain appropriate, taking into account any information obtained from the implementation or testing of the emergency arrangements.

PROTECTION STRATEGY

2.38. Paragraph 4.27 of GSR Part 7 [2] states:

"The government shall ensure that, on the basis of the hazards identified and the potential consequences of a nuclear or radiological emergency, protection strategies are developed, justified and optimized at the preparedness stage for taking protective actions and other response actions effectively in a nuclear or radiological emergency to achieve the goals of emergency response."

2.39. Paragraph 4.30 of GSR Part 7 [2] states that the government "shall ensure that interested parties are involved and are consulted, as appropriate, in the development of the protection strategy." Interested parties include regulatory bodies, consignors and carriers.

2.40. Protective actions for a transport emergency should be consistent with those for other emergencies and should be based on a reference level described in terms of residual dose (see para. 1.24 of IAEA Safety Standards Series No. GSR Part 3, Radiation Protection and Safety of Radiation Sources: International Basic Safety Standards [11], and para. 4.28(2) of GSR Part 7 [2]) and on national generic criteria expressed in terms of projected dose or received dose (see para. 4.28(3) of GSR Part 7 [2]).

2.41. Operational intervention levels (OILs)[10] for a nuclear or radiological emergency are provided in GSG-2 [6]. However, OILs can only be used in conjunction with observables and indicators to initiate an emergency response. Exceeding an OIL should not be used as the sole basis for initiating an emergency response. The declaration of an emergency class[11] should be based on specific observable conditions. In some rare cases, such as the active monitoring of the temperature of consignments, emergency action levels (EALs) may also be used to initiate an emergency response.

2.42. In incidents during transport, measured dose rates in excess of the operational intervention levels should not be used as the sole justification for declaring an emergency class and triggering emergency response actions. When dose rate measurements show that operational intervention levels are exceeded, the dose rate measurements should be compared with the measurement results and transport index recorded at the beginning of the shipment and with other observables and indicators to help identify abnormal conditions that justify the triggering of appropriate emergency response actions.

---

[10] Operational intervention levels are "A set level of a measurable quantity that corresponds to a generic criterion" [2].

[11] An emergency class is "A set of conditions that warrant a similar immediate emergency response" [2]. In accordance with para. 5.14(e) of GSR Part 7 [2], a transport emergency is classified as "other nuclear or radiological emergency", while recognizing that in some States the emergency classes may differ from those specified in GSR Part 7 [2].

2.43. For radioactive material regularly transported within a State, the size of any inner cordoned off area[12] to be established in an emergency should be determined and should be based on the national hazard assessment.

2.44. For recurring international shipments, governments should, when practicable, harmonize protection strategies for similar postulated emergencies through agreements or working groups.

PLANS AND PROCEDURES

2.45. The national arrangements for emergency preparedness and response relating to transport should incorporate the responsibilities of domestic and regular foreign consignors and carriers, as applicable. The emergency arrangements of the carrier should be consistent with the national arrangements of all the States in which it operates.

2.46. Transport emergencies should be addressed in the national radiation emergency plan (see para. 3.21 of GS-G-2.1 [5]). This plan should include the results of the hazard assessment and either include or make reference to the protection strategy.

2.47. Consignors and carriers are required by para. 304 of the Transport Regulations [3] to have emergency arrangements in place. These arrangements should be commensurate with the result of the hazard assessment and consistent with the national radiation emergency plan of States in which radioactive material is transported.

2.48. Additional plans and procedures may be developed for specific shipments that occur so infrequently that they are not considered in the national hazard assessment. This will depend primarily on the material being transported. Any such additional plans and procedures should be consistent with the existing plans and procedures, to the extent practicable.

2.49. In accordance with para. 5.6 of GSR Part 7 [2], emergency arrangements are required to be coordinated and integrated with the arrangements for response

---

[12] An inner cordoned off area is "An area established by first responders in an emergency around a potential radiation hazard, within which protective actions and other emergency response actions are taken to protect first responders and members of the public from possible exposure and contamination" [2].

to a nuclear security event during the transport of radioactive material (see Refs [12, 13]).

2.50. All response organizations should ensure that their plans are consistent and compatible with those of other response organizations. A process to ensure that any changes to existing plans are communicated to affected organizations should be developed and implemented.

National plans

2.51. There does not need to be a separate national plan for a transport emergency. In some States, the national radiation emergency plan is part of the all-hazards national emergency plan [14].

2.52. The emergency arrangements relating to transport should address all the relevant topics for the national radiation emergency plan. These topics include the following:

(a) Planning basis and hazard assessment;
(b) Authorities, responsibilities, capabilities and roles of the organizations involved;
(c) Procedures for alerting and notifying key organizations and persons;
(d) Methods of providing public information, including warning and informing;
(e) Generic criteria, emergency action levels (if appropriate), operational intervention levels, and observables and indicators;
(f) Protective actions and other response actions for the protection of people, property and the environment;
(g) Protection of emergency workers[13] and helpers;
(h) Resources for medical support;
(i) Training, drills and exercise programmes;
(j) Procedures for reviewing and updating plans and procedures;
(k) Procedures for response actions involving the recovery of a package;
(l) Conditions for terminating the emergency;
(m) Analysis of the emergency and the emergency response, including implementing actions to address identified gaps.

---

[13] An emergency worker is "A person having specified duties as a worker in response to an emergency" [2].

Local plans

2.53. Local authorities that have a role in responding to a transport emergency should develop a plan to enable emergency response functions to be performed. These local emergency plans relating to the transport of radioactive material should address all the necessary topics, including the following:

(a) A list of emergency response facilities in the local area, consistent with the national hazard assessment and planning basis;
(b) Responsibilities, capabilities and roles of the organizations involved, including the allocation of tasks and responsibilities during the response, and the responsibility for management of local operations;
(c) Procedures for requesting information and support from the consignor and the carrier in order to effectively respond to the emergency;
(d) Procedures for alerting and notifying key organizations and persons, including the fire brigade, police, emergency medical services, radiological assessors and other experts;
(e) Provision of public information, including warning and informing the public, and links with the media [9];
(f) Procedures for response actions, including the means of communicating with organizations involved in the response, and the conditions for terminating an emergency [7];
(g) Resources for medical support and for managing the medical response;
(h) Procedures for training, drills and exercises (see paras 3.43–3.53);
(i) Maintenance of the emergency plan.

Consignor and carrier plans

2.54. At the operating organization level, emergency plans for responding to a nuclear or radiological emergency involving the transport of radioactive material should conform, as closely as possible, to the plans for emergencies involving the transport of other dangerous goods. The plans for an emergency involving the transport of radioactive material should be integrated with the plans for other emergencies and for conventional emergencies, as appropriate.

2.55. Consignors and carriers conducting international shipments should ensure that their emergency arrangements comply with the regulatory requirements of each State through which they conduct shipments.

2.56. The emergency arrangements of consignors and carriers should be regularly reviewed and updated, as necessary, within a defined frequency, to take into account experience gained from drills, exercises and actual emergencies. The arrangements should be modified, as appropriate, based on any updates to regulations governing the transport of dangerous goods, the national radiation emergency plan or the local emergency plan. The arrangements should also be modified to take into account changes in the nature of transport activities. In addition, the contact details of personnel and organizations should be kept up to date. To simplify the process, contact names and communication details — which might change more frequently than other information — may be included as an annex to the plan.

2.57. Consignors and carriers should develop plans, as appropriate, for emergencies during the transport of radioactive material, commensurate with their hazard assessment. The plans should be documented and retained by carriers as part of the transport documents. The plans should also be made available to competent authorities. The plans should include the following:

(a) A description of the shipments covered under the plan.
(b) The initiating events that can be envisaged.
(c) The responsibilities of organizations involved in the transport of radioactive material, such as consignors, carriers, in-transit facilities, package designers, package owners and other organizations involved during the preparedness stage or during an emergency response.
(d) The procedures for identifying an emergency and for notifying the public safety authorities of the emergency if the carrier is incapacitated or unavailable.
(e) The procedures for coordination with public safety authorities.
(f) Any technical support that can be provided, including equipment that can be deployed to the site area for the following purposes:
  (i) Measurement and assessment (e.g. of leaktightness, dose rates, contamination levels, meteorological data);
  (ii) Mitigation of the radiological consequences (e.g. the provision of additional shielding and tarpaulins, replacement of damaged components, recovery of contaminated items);
  (iii) Package recovery, including specific means (e.g. lifting equipment, trailers, tie-down system) and a strategy for the removal of damaged packages to an interim location.
(g) The likely response actions (including instructions from the consignor to the carrier and response organizations).
(h) The response procedures and time frames for their implementation.
(i) The means of communication, documentation and recording to be used in the emergency response.

(j)   Templates and checklists for the activities of the carrier during the emergency response.

(k)   A quality management programme for emergency preparedness and response.

(l)   A programme for training, drills and exercises (see paras 3.43–3.53).

2.58. The consignor and carrier plans for emergency response should cover all phases of the response to an emergency, from the initial response actions to the transition phase, which includes preparation for the timely resumption of normal social and economic activity [7].

2.59. Annex III to this Safety Guide contains a template for an emergency response plan for consignors and carriers.

TRANSNATIONAL EMERGENCY ARRANGEMENTS

2.60. Appropriate communication and coordination systems should be used by all response organizations when establishing emergency arrangements and when responding to a transport emergency, taking into account the possibility of different States being involved in the shipment and the emergency response. These systems should include the designation of emergency contact points and mechanisms for communication.

2.61. All response organizations involved in an emergency that might occur during international transport should be aware of the notification process for the relevant national and local authorities of the State where the incident occurs (e.g. the means of communication, the language to be used, the persons to be contacted). In particular, the consignor should be able to quickly contact the authorities concerned to provide information, advice and assessments as necessary.

2.62. Consignors and carriers operating internationally should take into account international conventions and agreements, as well as the national legislation and regulatory requirements of all the States in which they operate. The consignor should ensure that the transport documents are written in the languages specified in the applicable international conventions and agreements, and in the national regulations.

2.63. Consignors and carriers should make arrangements with organizations in other States, as appropriate, to ensure the efficiency and effectiveness of arrangements and to comply with national requirements, such as language requirements.

2.64. The development of emergency arrangements should take into account that the consequences of an emergency might cross national borders, even if the shipment does not. Arrangements should be put in place to harmonize protective actions and other response actions, to the extent practicable, across national borders.

2.65. If frequent or recurring international shipments are planned, States may consider establishing bilateral or multilateral agreements that cover emergency arrangements. These agreements may include arrangements for cross-border deployment of personnel and resources to ensure that appropriately qualified personnel are permitted to respond across borders. These agreements may also include provisions for the advance exchange of information, results, findings and instructions intended for the public.

3. PREPAREDNESS AND RESPONSE ELEMENTS FOR TRANSPORT EMERGENCIES

3.1. This section provides recommendations on planning and implementing an emergency response, including the issues that should be considered when developing emergency arrangements. Emergency preparedness for the transport of radioactive material should consider a broad range of scenarios. The range of postulated emergencies should be identified at the national level and be based on a hazard assessment, as described in paras 2.29–2.37, for the types of shipment transported within the national territory.

PREPAREDNESS STAGE

3.2. The development of emergency arrangements should be completed prior to transport, in accordance with the graded approach defined in GSR Part 7 [2].[14] These emergency arrangements should take into consideration the actions that

---

[14] Emergency preparedness and response arrangements apply only for amounts of radioactive material exceeding the exemption levels specified by the regulatory body [11] and are commensurate with the level of hazard [2]. Therefore, the transport of amounts of radioactive material that only slightly exceed the exemption levels will need only limited emergency arrangements.

need to be performed in the event of a transport emergency, as well as the resources needed for the emergency response.

3.3. Unique shipments different from those addressed in the national hazard assessment may necessitate specific emergency arrangements.

CONCEPT OF OPERATIONS

3.4. A concept of operations is a brief description of an ideal response to a postulated emergency, used to ensure that all the personnel and organizations involved in the development of emergency response capabilities share a common understanding [2].

3.5. The concept of operations can be used by emergency planners to develop or revise their response plans.

3.6. To achieve the goals of emergency response described in para. 3.2 of GSR Part 7 [2], the following objectives should be considered when responding to a transport emergency:

(a) To establish and control the site area;
(b) To identify the radioactive material involved and the associated radiation hazards;
(c) To mitigate the consequences (e.g. fight the fire, contain the spill);
(d) To restore the package(s) to a safe, secure and stable state;
(e) To recover the radioactive material, package(s) and conveyance;
(f) To reopen the transport route to normal activity, after any necessary decontamination;
(g) To manage any radioactive waste arising from the emergency.

3.7. The concept of operations will describe a series of actions; however, the sequence of these actions will depend on the emergency conditions. There may be only a short time between the initiating event and the progression of an emergency, and the situation might have deteriorated when the responders arrive at the site.

3.8. This Safety Guide focuses on hazards from radioactive material. In some cases, other hazards might be present at the site area, and these might be the primary factor in determining response actions. The concept of operations should be applied in the context of plans and procedures for other hazardous substances and dangerous goods. In some cases, such as the transport of uranium hexafluoride, the radioactive material also presents chemical hazards that may outweigh the radiological hazards.

EMERGENCY RESPONSE PHASE

Urgent response phase

3.9. Once an initiating event has occurred and an emergency class has been declared, thereby activating the emergency response, preplanned response procedures based on the concept of operations should be implemented to make the appropriate notifications and initiate the emergency response.

3.10. The initial response to a transport emergency should be based on observable criteria and other indicators of conditions on the scene. A conveyance crew that is at an incident involving the transport of radioactive material, and first responders arriving at the site area, should identify observable conditions that could indicate an emergency. Any observable indication that radiological hazards might be present should be acted on, and preplanned response procedures should be implemented. An emergency class should be declared if there is a visible loss of containment or shielding integrity or if a measurement taken by a qualified individual with an appropriate radiation monitoring instrument indicates that dose rates are higher than should be expected. Leaking liquids, gases or powders may indicate that package integrity has been compromised.

3.11. Damage to a package, overpack, tank or container of radioactive material does not necessarily mean that the interior packaging containing the radioactive material or providing shielding has been damaged or breached. Nevertheless, external damage to a package is an indication that an emergency response needs to be activated, and the package should be examined by qualified personnel.

3.12. Even when there is no visible indication of external damage, an emergency response should be activated if accident conditions are developing that might lead to serious damage to package functions (e.g. in the case of a fire that cannot be extinguished in a timely manner).

3.13. In addition to the notification points[15] for the emergency services, the consignor's notification point should be listed in the transport documents or identified through other means, such as national programmes. In the event that a conveyance crew is unable to make the initial notification, for example because of injury or death, first responders may make the notification.

3.14. Initial responding organizations (i.e. first responders or the carrier) should, without delaying notifications, carry out an initial assessment by considering the following observable criteria:

(a) Location of the emergency.
(b) Available information regarding the site area, including its geography (e.g. hilly terrain, plains), local population, infrastructure or special environmental concerns.
(c) Site area accessibility.
(d) Nature of the initiating event (e.g. collision, fire, submersion).
(e) Injuries to people.
(f) Meteorological conditions.
(g) Labels, markings (e.g. UN numbers, proper shipping names), placards or transport documents.
(h) Other dangerous goods or other hazards present in the immediate vicinity of the site area, such as the following:
    (i) Large quantities of flammable liquids or gases;
    (ii) Explosive material;
    (iii) Toxic or corrosive materials.
(i) Any indications that the containment of any of the packages has been breached or that the shielding has been compromised.
(j) Any indications that the emergency was initiated by a malicious act.

3.15. The emergency response should be coordinated between response organizations in accordance with preplanned emergency arrangements and based on the necessary level of emergency response. The initial notification is likely to be general in nature; if so, further assessment will be needed to determine the resources and technical expertise (e.g. in criticality safety) that are needed.

---

[15] A notification point is "A designated organization with which arrangements have been made to receive notification…and to initiate promptly the predetermined actions to activate a part of the emergency response" [2].

3.16. After the initial notification, notification points should activate the appropriate response organizations, including any experts needed to assess the situation (e.g. radiological assessors), either at the site area or remotely.

3.17. The scope of the Convention on Early Notification of a Nuclear Accident [1] includes emergencies involving the transport of nuclear fuel, radioactive waste and radioisotopes used for agricultural, industrial, medical, and related scientific and research purposes. If an incident has resulted, or is likely to result, in a transboundary radioactive release that could be of radiological safety significance for another State, the competent authority of the accident State identified under the convention is obligated to notify the IAEA and the potentially affected State(s).

3.18. Emergency workers should access and review the transport documents (if available), which provide information on the type and number of packages, the radionuclides and the activity levels present. These documents will help determine the extent of the emergency and the expertise needed to respond. Consignors, consignees and carriers should make arrangements so that transport documents can be promptly made available to response organizations, on request.

3.19. First responders and other response organizations should ensure the following:

(a) Saving lives and administering first aid is given the highest priority. First responders should be trained and given information on the precautions to take while performing duties in the presence of radioactive contamination or elevated dose rates, including the application and use of operational intervention levels.
(b) Mitigatory actions, such as firefighting, are not delayed by the presence of radiological hazards. Response actions such as suppressing fires and dealing with flammable, explosive and toxic materials may take priority over an assessment of package integrity.
(c) A unified command and control for emergency response is established under the all-hazards approach. For a transport emergency, the unified command and control may include the consignor and/or carrier.
(d) All relevant response organizations have been effectively activated and have established appropriate communication channels.

3.20. Emergency workers should organize the site area consistent with national emergency arrangements and the guidance provided in GS-G-2.1 [5]. While the procedures for organizing the site area should be preplanned, additional considerations, based on the initial assessment, may be taken into account to determine the exact definition and location of specific areas. With regard to the organization of the site area, the following considerations apply:

(a) Checkpoints and command posts should be established upwind of any damaged package(s) and outside of any areas that might be affected by a spill of radioactive material.
(b) Cordoned off areas should be established immediately, for the protection of the public and of emergency workers. These areas should encompass any packages, overpacks, tanks or freight containers of radioactive material that have been ejected from a conveyance as a result of an incident. This may involve creating multiple cordoned off areas or one larger cordoned off area.
(c) The correct positioning of the boundaries of cordoned off areas should be periodically verified and modified, through radiation measurements, as necessary.

3.21. Emergency workers should consider any material released from a package to be hazardous until it has been determined by a radiological assessor, or (in the case of other hazardous substances) another appropriate expert, that the material is not hazardous.

3.22. The package integrity might have been impaired even if there are no visible indications; consequently, all packages involved in an incident should initially be treated with caution until appropriate surveys have been performed by a qualified individual (e.g. a radiological assessor).

3.23. The radiological assessor should assess the status of the package(s), including the containment, shielding, heat dissipation and criticality safety, as applicable. The certificate of approval for package design, which defines the design of the package, may be used to help the radiological assessor evaluate the integrity of the package safety features. If firefighting agents containing water have been used in proximity to fissile material, the radiological assessor should include this information in the criticality safety assessment.

3.24. Instruments, equipment and other supplies needed for mitigatory actions should be identified and made available so that they can be used promptly in an emergency. Mitigatory actions for damaged packages include the following:

(a) Using plugs, tarpaulins and leaktight overpacks to contain leaks and prevent the spread of contamination;
(b) Using additional shielding, as necessary;

(c)   Allowing packages to cool if they have been involved in a fire or if the heat dissipation system is damaged;

(d)   Recovering dispersed fissile material and placing it in special containers that have a safe geometry[16] and are watertight, ensuring that there is appropriate spacing between groups of packages containing fissile material.

3.25. Depending on the location of the emergency and on operational considerations at the site area, damaged packages may be transferred to an acceptable interim location (see para. 511 of the Transport Regulations [3]), following an assessment by qualified personnel (e.g. a radiological assessor) and with due precautions. However, any such packages should not be permitted to be forwarded to any other location until they have been repaired, reconditioned and decontaminated, as appropriate, by qualified personnel.

3.26. Radiation monitoring should be conducted as soon as possible during the emergency response to confirm the presence or absence of radiological consequences caused by the initiating event. The type of instrumentation selected should be based on the radionuclides expected to be present.

3.27. The results of dose rate measurements can be compared against the transport index to determine whether the package shielding has been damaged. If the dose rate exceeds 100 µSv/h (0.1 mSv/h) at a distance of more than 1 m from a single package containing radioactive material, it is likely that the package shielding has been compromised and operational intervention levels should be applied. This will not necessarily apply to shipments under exclusive use or to shipments where multiple packages are present. In such cases, other observables and indicators should be used.

3.28. During the urgent response phase, a thorough assessment of the radiological conditions in the site area might not be practicable. A full assessment of the situation could be an extended process, especially in cases involving the contamination of persons, objects and the environment.

3.29. Information confirming the loss of shielding or the release of radioactive material from a package, overpack, tank or freight container will typically only be obtained when radiation monitoring equipment is available. Any emergency workers who are expected to use radiation monitoring equipment — whether first

---

[16] A container with a safe geometry is one whose dimensions and shape are such that a criticality event cannot occur even with all other parameters at their worst credible conditions.

responders or other emergency workers — should be trained on how to use such equipment to measure dose rates and contamination, as appropriate.

3.30. Additional protective actions may need to be considered in the event of a loss of containment or shielding of the package(s). Such actions include the following:

(a) Control of access to and exit from the site area;
(b) Protective actions within and around the site area (e.g. sheltering, evacuation);
(c) Decontamination of persons;
(d) Actions for protecting the food chain and water supply;
(e) Protection of the local drainage system and drainage area.

3.31. Emergency plans and procedures should address how the news media and the public will be provided with information. For any transport emergency, concerted efforts should be made to keep the news media and the public well informed of the potential hazards and of what is being done to ensure public safety and to protect the environment. The public should be made aware of any protective actions that are recommended and the efforts that are being made for the transition of the site area back to its original condition. There should be no delays in providing this information, especially if delays might jeopardize the effectiveness of protective actions [9].

3.32. To minimize the risk of conflicting statements being issued to the news media, the responsibility of communicating with news media representatives should be coordinated between the relevant parties [9].

Early response phase

3.33. Prompt and continual assessment of radiological hazards and related hazards should be carried out by the radiological assessor to inform emergency responders and decision makers, in order to achieve the following objectives:

(a) To prevent the escalation of the emergency;
(b) To reduce the potential for, and to mitigate the consequences of, a radioactive release;
(c) To ensure that protection and safety are optimized during the response;
(d) To identify and obtain any additional expert support needed (e.g. chemical hazard assessors);
(e) To return the site area to a safe and stable state.

3.34. Depending on the type of consignment, the mode of transport and the severity of the incident, response organizations should consider whether early protective actions are necessary. These can include restrictions on the consumption of contaminated food, milk and drinking water, and on trade in commodities other than food (i.e. in cases where operational intervention levels might be exceeded).

3.35. If the drinking water supply is suspected of being contaminated as a result of the incident, it should be tested for contaminants. Similarly, in an emergency in or near a waterway where there is suspicion that radioactive material might have been released, the water should be tested for possible contamination.

3.36. Radiation monitoring should be performed during the emergency response to ensure that any protective actions and other response actions remain valid (or else are adjusted according to changing circumstances), and that the public is protected, based on operational intervention levels and other observables and indicators.

TRANSITION PHASE

3.37. The termination of a transport emergency might necessitate transition to either a planned exposure situation or an existing exposure situation, depending on the circumstances [2, 7, 11].

3.38. As stated in para. 2.1 of GSG-11 [7] (footnote omitted), "the transition phase commences as early as possible once the source has been brought under control and the situation is stable; the transition phase ends when all the necessary prerequisites for terminating the emergency have been met."

3.39. The transition from an emergency exposure situation will be subject to confirmation that certain prerequisites (i.e. in terms of the radiation hazard) have been fulfilled in the site area. This confirmation should be provided by the consignor, although specific arrangements may be made during the preparedness stage for the carrier, consignee or other organization to manage these actions. For a transport emergency, the site area is in the public domain; consequently, the relevant competent authorities and other relevant authorities will be involved in the final determination.

3.40. Consistent with the recommendations provided in GSG-11 [7], the following aspects should be considered by the consignor and response organizations during a transport emergency to help determine whether the necessary prerequisites for termination of the emergency have been met:

(a) Whether all radioactive material and all packages have been brought under control and are in a safe and stable condition. In some extreme situations, such as foundering in deep water, it might not be feasible to recover the packages. In such situations, an assessment of safety and stability should still be conducted, and a decision should be taken about whether to attempt recovery of the packages.

(b) Whether appropriate interim locations to receive the recovered items have been identified.

(c) Whether the movement of all packages and radioactive material from the site, including the appropriate transport documents, has been prepared; whether any required authorization has been applied for; and whether any necessary logistical arrangements have been made.

3.41. When assessing the situation, the consignor, in consultation with technical experts (e.g. package designers), should assess the likely development of the situation in the future. This may include, for example, corrosion of a containment system after a package has been submerged for an extended period.

3.42. In some cases, contamination levels might be high enough to warrant specific actions before terminating the emergency. Several decontamination and restoration methods might be employed, as appropriate, during the transition phase, including the following:

(a) Washing or vacuum sweeping roads and other objects and surfaces. This can be done with firefighting or industrial equipment. The water should be collected and disposed of safely.

(b) Washing and cleaning hard surfaces and equipment using water and appropriate detergents or other chemicals, and safely disposing of the liquids collected.

(c) Fixing contaminants using paints, strippable coatings, and paving materials such as asphalt. Depending on the nature of the radionuclides involved, the fixing agent may be removed after it has solidified, or it may be left in place.

(d) Removing or resurfacing contaminated road surfaces and/or removing contaminated soil.

TRAINING, DRILLS AND EXERCISES

3.43. The government is required to ensure that all relevant personnel who are likely to be involved in an emergency response receive training at an appropriate level (Requirement 25 of GSR Part 7 [2]). This training should be based on an assessment of the types of radioactive material transported in the region. Training programmes should be established for first responders, radiological assessors and other response organizations, in accordance with their response roles and functions. These programmes should include training on how to recognize and declare an emergency class.

3.44. Responders who would be arriving first at the site area should receive training that, at a minimum, will enable them to recognize and identify a nuclear or radiological emergency, implement preplanned protective actions, use personal protective equipment and notify the appropriate authorities.

3.45. In accordance with para. 313(c) of the Transport Regulations [3], persons engaged in the transport of radioactive material are required to receive additional training commensurate with their responsibilities in the event of a transport emergency.

3.46. All training should include information on the implementation of, and communication within, a unified command and control structure.

3.47. In accordance with para. 315 of the Transport Regulations [3] and para. 6.28 of GSR Part 7 [2], provision is required to be made for periodic refresher training ('retraining') to maintain the proficiency of personnel involved in the emergency response.

3.48. Relevant emergency response personnel are required to participate in drills and exercises in accordance with para. 6.31 of GSR Part 7 [2]. Drills are more limited in scope than exercises (see para. 3.50) and should be developed to maintain the skills of response personnel. Drills for shipments that have little or no potential to cause adverse radiological consequences should nevertheless be performed to test, at a minimum, the notification procedures and channels, as well as the procedures for verification of the integrity of the package and for repackaging.

3.49. Exercise programmes are required to be developed in accordance with para. 6.30 of GSR Part 7 [2]. Such programmes should be developed and implemented to ensure that scenarios involving shipments requiring a sizeable and resource intensive emergency response component are tested on a regular basis. These shipments might have the potential to exceed accident conditions of transport. Such exercise programmes should be designed to test all organizational interfaces, should be based on a graded approach and should include the participation of all the organizations concerned.

3.50. Exercises are required to be systematically evaluated in accordance with para. 6.33 of GSR Part 7 [2]. Emergency plans and procedures should be reviewed and — as needed — revised, based on exercise evaluation reports and as part of the quality management programme for emergency preparedness and response.

3.51. Radiological assessors and response organizations that have personnel with expertise in radiation protection or other relevant technical expertise and that may be called on for support and response in the event of a transport emergency need a detailed training programme. The personnel should be trained on the following on a regular basis, as appropriate for their assigned roles and responsibilities:

(a) Incident assessment techniques, using radiological monitoring instruments if appropriate;
(b) Criticality safety assessment;
(c) Determination and practical implementation of protective actions and other response actions;
(d) Use of protective clothing and equipment;
(e) Collection of contaminated material;
(f) Sealing techniques for leaking packages;
(g) Repacking of damaged packages;
(h) Dose estimation and dose reconstruction.

3.52. The representatives of the appropriate governmental authorities should receive training on the national emergency arrangements, the national transport safety regulations, and the roles and responsibilities of different authorities and organizations in responding to an emergency. These governmental authorities should have access to information about existing emergency response plans and the organizations that may be involved, as well as information about communication procedures and dealing with representatives of the news media.

3.53. A debriefing session should be performed as soon as possible after each drill, exercise and emergency. The emergency workers involved should take part in this debriefing session. Their reports and experiences should be documented and evaluated. The conclusions and lessons should be used to improve the emergency plans.

4. CONSIDERATIONS FOR MODES OF TRANSPORT

4.1. This section provides additional recommendations for specific modes of transport. These recommendations supplement the concept of operations described in paras 3.4–3.8. The recommendations in this section relate to safety; additional considerations may be necessary for nuclear security aspects, as described in Section 5.

TRANSPORT BY ROAD

4.2. The majority of shipments worldwide are conducted by road and these carry many types and quantities of radioactive material. Where limited emergency response resources are available, States may place restrictions on the transport of radioactive material through specific areas, such as areas with bridges, tunnels or seasonal routes. States may also completely close a transport route to shipments of radioactive material if an emergency response would not be feasible. They may also identify approved routes for the transport of radioactive material.

4.3. States may implement specific national requirements for planning certain types of shipment, based on the type and activity of radioactive material being transported, in accordance with a graded approach. These national requirements may also take into account the physical aspects of shipments, for example large or heavy consignments. Such national requirements may include route planning to avoid certain populated areas or essential infrastructure or restricting the transport of radioactive material to specific time periods.

4.4. A road transport emergency might occur in very close proximity to members of the public and may present unique challenges for initial response actions aimed at establishing the site area. Governments and first responders should be prepared for a road transport emergency anywhere in their territory, unless specific restrictions are in place, as described in paras 4.2 and 4.3.

4.5. The response measures for an emergency in an urban area might be different from those for an emergency in a rural area. The possible reasons for this include differences in the following areas:

(a) The availability of emergency response resources, including specially trained response teams and radiological assessors;
(b) Available communications systems and coverage areas;

(c) The number of people in the vicinity of the emergency;

(d) The surrounding environment, terrain and geography;

(e) Social and economic activity in the area (including the site area).

Areas with buildings for special populations, such as schools, nursing homes and hospitals, should be given special consideration in the protection strategy (see paras 2.38–2.44).

4.6. A road transport emergency might result in the blocking or temporary closure of the road, causing traffic congestion. This might hinder response actions, such as the arrival of emergency workers and the recovery of the radioactive material and package; it might also increase the number of people potentially affected by the emergency (e.g. occupants of vehicles blocked on the road).

TRANSPORT BY RAIL

4.7. Many types and quantities of radioactive material are transported by rail. In many cases, shipments are sent by rail due to size, weight or other operational considerations. Shipments by rail often involve large quantities of radioactive material that would otherwise necessitate multiple shipments by road; this should be considered in the hazard assessment.

4.8. Many of the considerations for emergency response to a road transport emergency (see paras 4.2–4.6) are also generally applicable to rail transport.

4.9. Trains can consist of large numbers of rail vehicles (i.e. railroad cars and railway wagons). This, together with other operational factors, might cause delays in identifying abnormal conditions, such as leaks or fires, that could result in an emergency.

4.10. Consignments of radioactive material are sometimes transported by trains that are dedicated to this cargo only. When such trains are involved in a transport emergency, several rail vehicles might be damaged, making the emergency response more complex.

4.11. Rail shipments may include a combination of radioactive material and other dangerous goods. At the preparedness stage, response organizations should consult with the competent authorities for transport to determine the possibility of encountering other dangerous goods during an emergency involving the transport of radioactive material by rail.

4.12. The railway authority will be directly involved in the emergency response. This may be a national (i.e. governmental) authority or a private company.

4.13. Carriers for rail transport should have established communication networks and procedures for notifying incidents and for activating an emergency response. However, they might have a limited ability to take initial response actions or mitigatory actions immediately after an emergency class is declared.

4.14. Accessing the site area of a rail transport emergency can be difficult if there is limited or no road access for response organizations. Furthermore, rail transport can traverse remote locations with difficult terrain. These factors should be taken into consideration in emergency arrangements, by making appropriate provisions for possible delays to initial response actions and the arrival on the site of emergency response organizations.

4.15. The location of the site area and the location of any affected consignments within the area might cause difficulties in restoring damaged packages to a stable state and removing them from the site area. This might result in an extended emergency response. Specialized equipment for operating on railways may be necessary to conduct the emergency response safely. This could include specially equipped rail vehicles with cranes, pumps and other safety equipment.

4.16. Compared with other modes of transport, rail transport entails a lower possibility of identifying short detours or partially reopening the transport route to minimize the impact of the emergency on the local population and on normal economic activity.

TRANSPORT BY SEA

4.17. Many types and quantities of radioactive material are shipped by maritime transport. This includes transport in international waters, ports or harbours, and the territorial seas and contiguous zones of States [15]. An emergency could be confined to the vessel or involve the release of radioactive material into the water.

4.18. For an emergency occurring in a port or harbour, specialized emergency response teams may be readily available. These port and harbour teams are usually trained to respond to marine related emergencies involving dangerous goods, and they may be needed during an emergency involving the transport of radioactive material. These teams should be provided with the appropriate level of training (see paras 3.43–3.53). The recommendations of the International Maritime

Organization [16] should be taken into account by the appropriate authorities, port operators, relevant cargo interests, emergency services and all others concerned.

4.19. The location of the emergency might be remote, such that the crew of the vessel are the only personnel available to deal with the emergency. Therefore, training should be provided to crews of ships carrying radioactive material on how to determine when to declare an emergency class and on the notification procedures to be followed to obtain quick and reliable information about the initial response actions to be taken. The crew should know that this information might be in the form of advice given via radio or other communication means, based on information gathered on board the vessel. In this context, the crew should comply with the guidance in Refs [17, 18]. Reference [17] provides specific instructions for fire and spillage emergencies on board a ship involving packaged dangerous goods, including radioactive material. Reference [18] gives general information about how to diagnose, treat and prevent health problems in seafarers, including the effects of radiation exposure, with a focus on the first 48 hours after injury, and specific instructions on how to treat crew members who are affected.

4.20. Maritime shipments may include the transport of a combination of different classes of dangerous goods. The possibility of encountering other dangerous goods in an emergency involving radioactive material should be taken into account by the response organizations.

4.21. A cordoned off area on board a vessel should be established under the authority of the ship's master. In the event that the site area extends beyond the boundaries of the vessel to the open sea, the ship's master should communicate a warning to other vessels, for example a pan-pan or mayday message, given that establishing physical barriers is not possible. If the vessel is within a port or harbour, the ship's master should communicate a warning to the responsible authorities and coordinate actions with those authorities to establish and maintain the cordoned off area, as necessary.

4.22. If a seagoing vessel is seeking safe harbour during an emergency, the ship's master should, as soon as possible, communicate the current emergency situation and response actions to the responsible authorities in the port or harbour, or the relevant coastal authorities.

4.23. If an emergency involves the loss (or likely loss) of packaged radioactive material overboard into the sea, the ship's master should fully report the situation without delay to the nearest coastal State [18]. Any radioactive release to the atmosphere that might impact vessels at sea or in port should be reported to the NAVAREA coordinators[17] [19].

4.24. For an emergency involving a release of radioactive material into the sea, emergency workers may need access to expertise in marine dispersion modelling and in monitoring and sampling to determine whether to implement protective actions or other response actions. Access to such expertise may be needed in the urgent response phase to consider the effects of radioactive material being carried by marine currents. Expertise may also be needed in the early phase or transition phase owing to other factors, such as corrosion.

4.25. In an emergency involving a possible release of radioactive material into the water, protective actions relating to placing restrictions on normal activities (e.g. fishing) should be based on the following:

(a) An assessment of the package and the possible magnitude and duration of the release;
(b) The chemical form of the radioactive material and its reactivity to water;
(c) In some cases, maritime monitoring and sampling, as well as seafood sampling.

4.26. Retrieving sunken packages or vessels will involve specialized teams capable of maritime salvage operations. In some cases, retrieval of the package might not be justified from a radiation protection perspective. This decision should be based on the protection strategy put in place by the national government(s) responsible for the site area, or the flag State of the vessel in the event of an emergency in international waters. Packages sunk in shallow waters should be recovered unless doing so is either impossible or cannot be justified.

4.27. Emergencies at sea might not be covered in detail in the national radiation emergency plan. Accordingly, a ship's master should have information on which authorities in the ports of call, or in States, to contact in an emergency along the transport route. The maritime authorities with whom the master might be in contact during a voyage should also know whom to contact in an emergency so that, if the vessel needs to go into port, the emergency services will have been alerted in advance.

---

[17] For the purposes of the World-Wide Navigational Warning Service, a coordinated global service for the promulgation of navigational warnings, the world's oceans are divided into 21 geographical sea areas termed NAVAREAs (navigational areas). In each area, one designated country is responsible for disseminating navigational information.

4.28. Vessels subject to the International Code for the Safe Carriage of Packaged Irradiated Nuclear Fuel, Plutonium and High-Level Radioactive Wastes on Board Ships (the 'INF Code') [20] should have on board a shipboard emergency plan, developed in accordance with Ref. [21]. At a minimum, the plan should consist of the following:

(a) The procedure to be followed by the master or other persons having charge of the ship to report an incident involving INF cargo;
(b) The list of authorities or persons to be contacted in the event of an incident involving INF cargo;
(c) A detailed description of the action to be taken immediately by persons on board to prevent, reduce or control the release, and mitigate the consequences of the loss, of INF cargo following the incident;
(d) The procedures and points of contact on the ship for coordinating shipboard action with national and local authorities.

4.29. For an emergency on a vessel subject to the INF Code [20], coastal States may have relevant information provided through voluntary and confidential government-to-government communications [22]. Response organizations notified of an emergency within their territorial seas, or vessels requesting safe harbour, should check with their national competent authorities to see if such information is available.

4.30. In the event of a conventional emergency while in harbour (e.g. earthquake, tsunami warning), carriers with vessels subject to the INF Code should have criteria and procedures for emergency shore leave that are commensurate with the hazard assessment.

TRANSPORT BY INLAND WATERWAY

4.31. Inland waterway transport involves transport in proximity to land and on the landward side of the baseline of the territorial seas of a State. In the Transport Regulations [3], the requirements for inland waterway craft are defined separately from those for seagoing craft; as such, this mode of transport presents unique challenges for emergency response. Compared with seagoing vessels, the conveyance activity limits and transport index limits for inland waterway craft are generally lower.

4.32. Although inland waterway craft are generally smaller than seagoing vessels, many of the maritime transport emergency response considerations are generally applicable to inland waterway transport.

4.33. Compared with maritime transport, response organizations and equipment may be closer and more readily available during inland waterway transport, and this should be reflected in the emergency arrangements. However, any response to an emergency on a vessel presents common challenges compared with a response on land.

4.34. Transport through inland waterways occurs most often in national waterways. However, some inland waterways are designated as international waterways and have a unique legal status. Inland waterways may also serve as national borders, resulting in a transboundary emergency even when there are limited radiological consequences.

TRANSPORT BY AIR

4.35. Many types and quantities of radioactive material are shipped by air on both passenger aircraft and cargo aircraft. This includes the frequent transport of short lived radionuclides for medical applications. Much less frequent is the transport by air of high activity radioactive sources in Type C packages. An emergency involving the transport of radioactive material by air could occur either at airports or at locations along the route of the aircraft.

4.36. Recognizing a nuclear or radiological emergency during transport by air can be difficult, and the initial response actions will follow procedures for a conventional emergency. The pilot in command should provide information on any dangerous goods, including radioactive material, carried as cargo on the aircraft. If the pilot in command is incapacitated, the airline should provide information to the response organizations as soon as possible.

4.37. An emergency that is the result of a crash may involve an emergency response in a remote or inaccessible area. Aircraft crashes often involve strong deceleration forces and a high probability of fire compared to other modes of transport. The radioactive material could be scattered over a wide area and be difficult to locate and retrieve. Emergency workers should be aware of the possibility of high dose rates and airborne and surface contamination resulting from serious damage to packages and should take appropriate precautions and use personal protective equipment.

4.38. When attempting to locate and recover radioactive material, emergency workers should be aware that some of the packages and their contents might have physical and chemical characteristics that are different from the characteristics before the crash. The particle size of the dispersed radioactive material might vary depending on the forces and temperatures involved in the initiating event.

4.39. Type C packages are designed to withstand most aircraft crashes [23, 24]. For other types of package, emergency workers at the site of an aircraft crash should consider the possibility that the package has been damaged or destroyed and that its shielding has been lost. If the aircraft was carrying packages containing high activity radioactive sources, additional precautions should be taken by emergency workers to ensure the radiation protection of the public and emergency workers, as described in Section 3.

4.40. In the event of a crash of an aircraft carrying packages containing fissile material, the criticality hazard is reduced if the fissile material is spread over a large area. Nevertheless, an assessment should be carried out to confirm the absence of criticality hazards and to determine appropriate actions for maintaining this absence. Special care should be taken when collecting the fissile material.

4.41. Some aircraft use radioactive material as part of their construction, for example depleted uranium counterweights. These materials are not part of a consignment and are outside the scope of the Transport Regulations [3]; however, these materials may require some response actions in accordance with GSR Part 7 [2].

5. INTERFACE WITH NUCLEAR SECURITY

5.1. An emergency involving the transport of radioactive material might be initiated by a nuclear security event. This section provides considerations to be addressed in the management of an emergency response whenever it is suspected that a nuclear security event might have initiated the emergency. Even in an emergency that is not initiated by a nuclear security event, there may be a need to implement nuclear security measures to secure the radioactive material.

5.2. When developing emergency arrangements, as required by GSR Part 7 [2] and the Transport Regulations [3], operating organizations (consignors, carriers and consignees) should ensure that appropriate contingency plans for nuclear security are considered. The response to a transport emergency initiated by a nuclear security event should be integrated with the response to an emergency under a unified command and control system at the local, national, regional and international levels, as appropriate. More information can be found in GSR Part 7 [2] and in Ref. [25].

5.3. Paragraph 4.22 of GSR Part 7 [2] states: "The government shall ensure that the hazard assessment includes consideration of the results of threat assessments made for nuclear security purposes".

5.4. The following considerations should be taken into account with regard to a transport emergency initiated by a nuclear security event:

(a) Sabotage can lead to an emergency at the site area where the incident occurs. The site should be deemed a radiological crime scene. Therefore, the response to this scenario should include both emergency response actions and nuclear security measures. Reference [25] provides guidance on nuclear security measures, based on the nature and the activity of the radioactive material involved in the event.

(b) The unauthorized removal of radioactive material during transport might lead to an emergency at an unpredictable location. Response to this scenario is beyond the scope of this publication; however, the requirements established in GSR Part 7 [2] and the guidance provided in Refs [25, 26] can be used to provide input to the arrangements for responding to such an event.

5.5. Some shipments incorporate safety measures on the package (e.g. seals) and on the conveyance (e.g. tie-down requirements) that could help deter, detect or delay an adversary from gaining access to the package or the radioactive material.

5.6. Response organizations might face conflicting priorities when responding to a transport emergency that was initiated by a nuclear security event. For example, for nuclear security purposes, the integrity of a radiological crime scene needs to be maintained for criminal investigation and evidence collection. However, when necessary during the emergency response, life saving actions and mitigatory actions take priority. The final decision on prioritizing specific tasks and actions should be made within the unified command and control system, established and used in accordance with para. 5.7 of GSR Part 7 [2].

CONSIDERATIONS FOR THE EMERGENCY RESPONSE WHEN A NUCLEAR SECURITY EVENT IS CONFIRMED TO BE THE INITIATING EVENT

5.7. A State should establish a comprehensive all-hazards national emergency plan that includes response to a transport emergency in cooperation and coordination with the national response plan for a nuclear security event. As a generic consideration in the initial response, all emergency response actions should be undertaken considering the possibility of a nuclear security event.

5.8. The responses to an emergency and to a nuclear security event may be based on different approaches. Consequently, at the preparedness stage, considerations relating to nuclear security measures should be included in the unified command and control system (see para. 2.10). This will help address possible conflicts in advance.

5.9. The IAEA Nuclear Security Series contains guidance on nuclear security measures. This guidance includes the following:

(a) Radiological crime scene management [25]: The emergency site area associated with a nuclear security event may contain evidence of activities that indicate a criminal or unauthorized act involving radioactive material. Law enforcement operations and emergency response activities should be carried out simultaneously and in a coordinated manner, taking into consideration the need to protect emergency workers, helpers and the public. Actions for protecting persons, whether the actions are for radiation protection purposes or for responding to malicious acts, should take priority over other activities, such as collecting evidence, interviewing witnesses, taking photographic images and preparing written records of the scene.
(b) Forensic examination [27]: This includes traditional forensic examination conducted by law enforcement agencies and nuclear forensic examination conducted by special experts. If the emergency involves unknown radioactive material, nuclear forensic examination should be carried out to answer questions regarding the nature, history and origin of the radioactive material involved.
(c) Criminal investigation activities [25]: These activities are undertaken in accordance with national procedures for criminal investigations, which are aimed at obtaining evidence from individuals near the emergency site area who may have witnessed events leading up to, during or immediately following the emergency.

5.10. Effective, timely and clear communication within the government and with the news media and the public is essential, as described in previous sections. Taking nuclear security issues into consideration, provisions should be put in place for controlling sensitive information (e.g. certain information dealing with the law enforcement response and crime scene investigations) so that law enforcement is not impeded.

5.11. The capabilities and resources relating to nuclear security measures that need to be available (and integrated into the unified command and control system) as part of the response to a transport emergency include the following:

(a) Nuclear forensics support;
(b) Equipment for secure communication;
(c) Specialized equipment, such as explosives detectors or equipment for handling pyrophoric material, and personnel able to use it;
(d) Resources for delivering and analysing evidence.

Appendix I

CONSIDERATIONS FOR DEVELOPING A NATIONAL CAPABILITY

I.1. This appendix describes specific actions that a State should complete so that it can respond effectively to an emergency during the transport of radioactive material. The level of emergency arrangements and plans necessary should be derived from the conclusions of the hazard assessment (i.e. via the normal process of drawing up emergency plans). Difficulties can arise at different points in this process, for example owing to limited knowledge, practical experience or regulatory infrastructure in a State. The objective of this appendix is to draw attention to considerations relevant to addressing these issues.

ESTABLISHING THE COORDINATING ORGANIZATION AND THE NATIONAL POLICY

I.2. Developing a national capability involves extensive coordination between all the relevant ministries, agencies and organizations involved. It is a dynamic process (i.e. plans and procedures will need to be developed and revised throughout). The general role of the leading organization(s) should be consistent with the need to coordinate the contributions of all national organizations that will be involved in preparedness and response for an emergency, and with the need to integrate these organizations' contributions into a national all-hazards emergency management system.

CONDUCTING THE NATIONAL HAZARD ASSESSMENT

I.3. The national hazard assessment starts with identifying the different fundamental characteristics of the radioactive material transported within the State, before identifying the specific radioactive material that may transit through the State's territorial land or waters. The following list of facilities and activities involving the use or transport of radioactive sources may help identify potential consignors, carriers and consignees:

(a) Mining and separation and concentration plants (e.g. uranium ores and tailings, density gauges);
(b) Agricultural facilities and industrial buildings (e.g. density and moisture gauges, smoke detectors);

(c)   Industrial radiography companies;

(d)   Hospitals and laboratories (e.g. radiopharmaceuticals, gamma radiotherapy sources);

(e)   Nuclear installations (e.g. fuel fabrication facilities, research reactors, nuclear power plants and waste repositories);

(f)   In-transit facilities (e.g. ports, airports, rail terminals);

(g)   Facilities that generate radioactive waste, and disposal facilities;

(h)   Industrial facilities (e.g. irradiation facilities, nuclear gauges).

I.4.   After collecting the information described in para. I.3, a survey of the transport activities undertaken in the State should be carried out to determine the following:

(a)   The nature and frequency of shipments (classified in accordance with UN numbers);

(b)   The types and quantities of radioactive material currently transported;

(c)   The types of package in each type of consignment;

(d)   The primary routes used, and in the case of frequent shipments, representative routes of common shipments;

(e)   The locations within these routes with specific transport related risks (e.g. tunnels, bridges, mountains, seasonally damaged roads);

(f)   For each primary or representative route: the terrain, the local geographical conditions and the nearby population distribution;

(g)   Any existing nuclear security contingency plans.

A systematic assessment of this information will help determine the potential nature and magnitude of the nuclear or radiological hazards that might be associated with a transport emergency. The result of this analysis can then be used to implement a graded approach to emergency arrangements, commensurate with the potential nature and magnitude of each hazard.

DEVELOPING THE PLANNING BASIS

I.5.   Once the hazard assessment has been completed, it is necessary to gather more information for the planning process. This may include information on the following:

(a)   Laws or regulations establishing criteria for the protection of emergency workers, helpers and the public;

(b) International agreements governing international trade or the response to an emergency (e.g. the Convention on Early Notification of a Nuclear Accident and the Convention on Assistance in the Case of a Nuclear Accident or Radiological Emergency [1] or regional transport agreements);

(c) Any bilateral and multilateral emergency arrangements;

(d) Consignors, carriers and in-transit facilities;

(e) National coordinating mechanisms for planning the response to a nuclear or radiological emergency and for planning the response to a conventional emergency;

(f) Procedures for notifying other States and requesting international assistance;

(g) Arrangements for making decisions on protective actions and other response actions and arrangements for implementing those actions;

(h) Arrangements for providing emergency services support;

(i) Arrangements for providing a response to criminal activities;

(j) Off-site monitoring and laboratory analysis resources;

(k) Means of communication available for decision makers;

(l) Means of communication available to alert and inform the public;

(m) The assistance available from other operating organizations that could provide support in the response;

(n) Off-site environmental conditions (e.g. severe conditions that could result in an emergency).

DEVELOPING A CONCEPT OF OPERATIONS AND ASSIGNING
DETAILED RESPONSIBILITIES

I.6.   A basic concept of operations (see paras 3.4–3.8) needs to be developed describing the response process.

I.7.   On the basis of this concept of operations, the roles and responsibilities of each organization involved in emergency preparedness and response need to be determined and assigned. A list of key responsibilities and tasks should be assigned: for emergency management operations, for the initial response (identifying, notifying and activating), and for all other response actions (mitigatory actions, urgent protective actions, early response actions and other response actions).

I.8.   The assignment of responsibilities is an interactive process; it should be carried out in consultation with each organization and should take into account the capabilities of each organization. The organizations to which roles and responsibilities are assigned should agree to the assignments. The assignment of responsibilities should be based on the related laws and regulations.

WRITING THE PLANS AND PROCEDURES AND INTEGRATING THEM INTO THE NATIONAL RADIATION EMERGENCY PLAN

I.9. Developing the plan for response to an emergency that involves the transport of radioactive material should not be separated from developing the national radiation emergency plan.

I.10. The plan should enable preparation for representative emergencies derived from the hazard assessment by identifying appropriate response mechanisms to a variety of potential hazards that might arise during the transport of radioactive material. The plan should provide an incident management structure to guide response activities and should outline the necessary resources, personnel and logistics needed for a prompt, coordinated and rational approach to a broad range of transport incidents.

I.11. The plan should contain sufficient detail but be flexible enough to enable those involved in the response to carry out their duties effectively. All response organizations in the plan should be given an opportunity to review the plan.

IMPLEMENTING DETAILED ARRANGEMENTS

I.12. Each organization that has a role in implementing the national emergency response plan should develop its capabilities relative to the functional and infrastructural requirements in GSR Part 7 [2]. These arrangements include plans, procedures, organizational structure, staffing, facilities, equipment and training. These arrangements need to be addressed by the operating organization (consignor, carrier or consignee, as appropriate), the local authorities and the national authorities.

I.13. Through the national coordinating mechanism, a coordinating committee should be assigned the responsibility of assisting with the implementation of the arrangements described in para. I.12. The duties of this coordinating committee should include the following:

(a) To prepare the criteria and the schedule for the development of plans and procedures for each organization involved;

(b) To provide assistance to individual organizations in the development of plans and procedures to ensure compatibility and completeness of the planning process;

(c) To organize periodic meetings between key representatives to encourage coordination;

(d) To verify that progress is consistent with the schedule or, if necessary, update the schedule.

TESTING THE NATIONAL CAPABILITY

I.14.   By itself, a finished national emergency response plan does not ensure readiness. Drills and exercises should be conducted to test and demonstrate the adequacy of the arrangements. The numerous interactions that will take place between response organizations (including the interfaces with nuclear security) warrant extensive training and regular exercises so that all parties concerned are adequately prepared. The plans and procedures, and the capabilities of the national infrastructure, should be reviewed and revised to take into account the results of the evaluation of drills and exercises.

Appendix II

TYPES OF EVENT THAT MIGHT LEAD TO
A TRANSPORT EMERGENCY

II.1. Packages containing radioactive material are transported worldwide via road, rail, inland waterway, sea and air. Incidents can occur while these packages are being transported, handled (loading and unloading) or stored temporarily in transit. This appendix provides some information on the types of event that might occur during transport and initiate a nuclear or radiological emergency.

GENERAL CONSIDERATIONS FOR AN EMERGENCY

II.2. The loadings to which packages are subjected during different types of transport accident vary considerably. When performing the national hazard assessment, the State will identify emergency scenarios and determine their potential consequences as a basis for establishing arrangements for emergency preparedness and response. In determining potential radiological consequences, the range of potential initiating events and the parameters to be considered for all the scenarios identified is very wide. The assessment may be simplified by considering only those scenarios that would have the most severe consequences. To determine the parameters associated with such scenarios, States may use data from international modal databases, such as the International Maritime Organization's Global Integrated Shipping Information System (GISIS) for maritime transport events and the International Civil Aviation Organization's Accident/Incident Data Reporting (ADREP) system for air transport events. There may also be other sources of data available to States.

II.3. In setting parameters, scenarios that might occur on different modes of transport, such as fires of a long duration (i.e. longer than the thermal test in the Transport Regulations [3]), should be considered. For example, tunnel environments during road or rail transport should be considered if consignments of radioactive material are permitted to pass through them. Additionally, packages might be impacted by other objects during transport, for example if heavy objects are dropped onto a package at a seaport, airport or other facility where heavy objects are frequently moved.

TYPES OF EVENT DURING ROAD TRANSPORT

II.4. The main types of road accident that should be considered for emergency planning purposes are the following:

(a) Collision;
(b) Fire or explosion;
(c) Immersion or flooding;
(d) Loss of load or spillage.

These might occur as a single event or as a sequence of events, but commonly the initiating event of a road accident is a collision. This occurs when a vehicle collides with another vehicle or a stationary object (e.g. tree, pole, wall), possibly resulting in injury or death and in damage to property. A number of factors contribute to the risk of road vehicle collisions, such as vehicle design, speed of operation, driver skill and behaviour, defective roads, traffic and weather conditions.

Likelihood of an initiating event

II.5. Road accidents are the most frequent type of accident involving the transport of radioactive material. These accidents are mainly the result of vehicle collisions. Such accidents could result in package damage and, depending on the severity of the accident and the types of package being transported, the spread of contamination in the immediate area.

II.6. Fires and explosions are likely to be the most severe scenarios for planning purposes because they entail a larger potential impact on the public in the vicinity than other scenarios due to the loss of containment of the package and the dispersion of radioactive material.

TYPES OF EVENT DURING RAIL TRANSPORT

II.7. The types of rail accident are similar to the types of road accident and might involve single events or a sequence of events, as follows:

(a) Collision;
(b) Fire or explosion;
(c) Loss of load or spillage.

Rail accidents occur when trains travelling on the same tracks collide; when trains derail because of technical faults in the rolling stock, the rails or the systems for securing the rail vehicles; because of excess speed; or because of landslides, avalanches or objects obstructing the rails, potentially caused by deliberate actions such as terrorist attacks.

Likelihood of an initiating event

II.8. When rail vehicles are transporting radioactive material, impact due to collision or derailment could lead to package damage. Trains often carry large quantities of goods, and serious rail accidents can damage several rail vehicles at once, potentially resulting in contamination of a larger area than might result from a road transport accident.

II.9. Fire and explosion are likely to be the most severe scenarios for planning purposes because they entail a larger potential impact on the public in the vicinity than other scenarios due to the loss of containment of the package and the dispersion of radioactive material.

TYPES OF EVENT DURING MARITIME TRANSPORT

II.10. Maritime accidents can be divided into the following categories:

(a) Collision;
(b) Grounding;
(c) Contact;
(d) Fire or explosion;
(e) Hull failure;
(f) Loss of control;
(g) Ship or equipment damage;
(h) Capsizing or listing;
(i) Flooding or foundering;
(j) Missing ship;
(k) Cargo damage due to heavy rolling;
(l) Cargo damage during loading or unloading.

These events might occur alone or in combination. Events that involve casualties need to be reported to the International Maritime Organization under the International Convention for the Safety of Life at Sea [28] and the International

Convention for the Prevention of Pollution from Ships [29] through GISIS. This information is available to International Maritime Organization Member States.

Likelihood of an initiating event

II.11. Collision, grounding and contact are the most common events. For planning purposes, fires and explosions are likely to be the most severe scenarios. The other types of event listed in para. II.10 vary in likelihood depending on the types of vessel involved.

TYPES OF EVENT DURING AIR TRANSPORT

II.12. Air accidents can be divided into the following categories:

(a) Ground impact;
(b) In-flight impact or collision;
(c) Ground fire (during ground operation, post-impact, during an aborted take-off or following landing);
(d) In-flight fire;
(e) In-flight explosion;
(f) Immersion;
(g) Events during loading and unloading at airports.

II.13. More information on air accidents can be found in the International Civil Aviation Organization's ADREP system, which is available to that organization's Member States. Air accidents can be of natural, technical or human origin; possible causes include severe weather, mechanical issues, negligence, pilot error or terrorist attack.

II.14. The majority of air accidents are single event airplane accidents resulting in ground impact and post-impact fire.

Likelihood of an initiating event

II.15. The frequency of air transport accidents is low compared with that of other modes of transport. If an accident involving an aircraft occurs, various accident conditions can be generated, imposing stresses on packages containing radioactive material. For planning purposes, the most severe consequences are likely to involve a high impact accident or a fire where the cargo includes

Type B(U) or Type B(M) packages.[18] Such accidents could result in extensive damage to the package shielding and loss of containment, resulting in significant dose rates in the vicinity of the package and the dispersion of radioactive material.

ADDITIONAL CONSIDERATIONS

II.16. Additional considerations for developing the postulated scenarios relating to specific events during transport can be found in Annex IV.

---

[18] For transport by air, the content of a Type B(U) or Type B(M) package is limited to $3000A_1$ or 10^5A_2, whichever is the lower, for special form radioactive material; or $3000A_2$ for all other radioactive material (see para. 43 of the Transport Regulations [3]).

REFERENCES

[1]   INTERNATIONAL ATOMIC ENERGY AGENCY, Convention on Early Notification of a Nuclear Accident and Convention on Assistance in the Case of a Nuclear Accident or Radiological Emergency, Legal Series No. 14, IAEA, Vienna (1987).

[2]   FOOD AND AGRICULTURE ORGANIZATION OF THE UNITED NATIONS, INTERNATIONAL ATOMIC ENERGY AGENCY, INTERNATIONAL CIVIL AVIATION ORGANIZATION, INTERNATIONAL LABOUR ORGANIZATION, INTERNATIONAL MARITIME ORGANIZATION, INTERPOL, OECD NUCLEAR ENERGY AGENCY, PAN AMERICAN HEALTH ORGANIZATION, PREPARATORY COMMISSION FOR THE COMPREHENSIVE NUCLEAR-TEST-BAN TREATY ORGANIZATION, UNITED NATIONS ENVIRONMENT PROGRAMME, UNITED NATIONS OFFICE FOR THE COORDINATION OF HUMANITARIAN AFFAIRS, WORLD HEALTH ORGANIZATION, WORLD METEOROLOGICAL ORGANIZATION, Preparedness and Response for a Nuclear or Radiological Emergency, IAEA Safety Standards Series No. GSR Part 7, IAEA, Vienna (2015).

[3]   INTERNATIONAL ATOMIC ENERGY AGENCY, Regulations for the Safe Transport of Radioactive Material, 2018 Edition, IAEA Safety Standards Series No. SSR-6 (Rev. 1), IAEA, Vienna (2018).

[4]   INTERNATIONAL ATOMIC ENERGY AGENCY, IAEA Safety Glossary: Terminology Used in Nuclear Safety and Radiation Protection, 2018 Edition, IAEA, Vienna (2019).

[5]   FOOD AND AGRICULTURE ORGANIZATION OF THE UNITED NATIONS, INTERNATIONAL ATOMIC ENERGY AGENCY, INTERNATIONAL LABOUR OFFICE, PAN AMERICAN HEALTH ORGANIZATION, UNITED NATIONS OFFICE FOR THE COORDINATION OF HUMANITARIAN AFFAIRS, WORLD HEALTH ORGANIZATION, Arrangements for Preparedness for a Nuclear or Radiological Emergency, IAEA Safety Standards Series No. GS-G-2.1, IAEA, Vienna (2007).

[6]   FOOD AND AGRICULTURE ORGANIZATION OF THE UNITED NATIONS, INTERNATIONAL ATOMIC ENERGY AGENCY, INTERNATIONAL LABOUR OFFICE, PAN AMERICAN HEALTH ORGANIZATION, WORLD HEALTH ORGANIZATION, Criteria for Use in Preparedness and Response for a Nuclear or Radiological Emergency, IAEA Safety Standards Series No. GSG-2, IAEA, Vienna (2011).

[7]   FOOD AND AGRICULTURE ORGANIZATION OF THE UNITED NATIONS, INTERNATIONAL ATOMIC ENERGY AGENCY, INTERNATIONAL CIVIL AVIATION ORGANIZATION, INTERNATIONAL LABOUR OFFICE, INTERNATIONAL MARITIME ORGANIZATION, INTERPOL, OECD NUCLEAR ENERGY AGENCY, UNITED NATIONS OFFICE FOR THE COORDINATION OF HUMANITARIAN AFFAIRS, WORLD HEALTH ORGANIZATION, WORLD METEOROLOGICAL ORGANIZATION, Arrangements for the Termination of a Nuclear or Radiological Emergency, IAEA Safety Standards Series No. GSG-11, IAEA, Vienna (2018).

[8]   INTERNATIONAL ATOMIC ENERGY AGENCY, Governmental, Legal and Regulatory Framework for Safety, IAEA Safety Standards Series No. GSR Part 1 (Rev. 1), IAEA, Vienna (2016).

[9]   FOOD AND AGRICULTURE ORGANIZATION OF THE UNITED NATIONS, INTERNATIONAL ATOMIC ENERGY AGENCY, INTERNATIONAL CIVIL AVIATION ORGANIZATION, INTERPOL, PREPARATORY COMMISSION FOR THE COMPREHENSIVE NUCLEAR TEST BAN TREATY ORGANIZATION, UNITED NATIONS OFFICE FOR OUTER SPACE AFFAIRS, Arrangements for Public Communication in Preparedness and Response for a Nuclear or Radiological Emergency, IAEA Safety Standards Series No. GSG-14, IAEA, Vienna (2020).

[10]   INTERNATIONAL ATOMIC ENERGY AGENCY, Dangerous Quantities of Radioactive Material (D-Values), EPR-D-Values 2006, IAEA, Vienna (2006).

[11]   EUROPEAN COMMISSION, FOOD AND AGRICULTURE ORGANIZATION OF THE UNITED NATIONS, INTERNATIONAL ATOMIC ENERGY AGENCY, INTERNATIONAL LABOUR ORGANIZATION, OECD NUCLEAR ENERGY AGENCY, PAN AMERICAN HEALTH ORGANIZATION, UNITED NATIONS ENVIRONMENT PROGRAMME, WORLD HEALTH ORGANIZATION, Radiation Protection and Safety of Radiation Sources: International Basic Safety Standards, IAEA Safety Standards Series No. GSR Part 3, IAEA, Vienna (2014).

[12]   INTERNATIONAL ATOMIC ENERGY AGENCY, Security of Radioactive Material in Transport, IAEA Nuclear Security Series No. 9-G (Rev. 1), IAEA, Vienna (2020).

[13]   INTERNATIONAL ATOMIC ENERGY AGENCY, Security of Nuclear Material in Transport, IAEA Nuclear Security Series No. 26-G, IAEA, Vienna (2015).

[14]   INTERNATIONAL ATOMIC ENERGY AGENCY, Method for Developing Arrangements for Response to a Nuclear or Radiological Emergency, EPR-METHOD 2003, IAEA, Vienna (2003).

[15]   UNITED NATIONS, United Nations Convention on the Law of the Sea (with annexes, final act and procès-verbaux of rectification of the final act dated 3 March 1986 and 26 July 1993), concluded at Montego Bay on 10 December 1982, United Nations Treaty Series, Vol. 1833 (1994) 397–581.

[16]   INTERNATIONAL MARITIME ORGANIZATION, Revised Recommendations on the Safe Transport of Dangerous Cargoes and Related Activities in Port Areas, 2007 Edition, MSC.1/Circ.1216, IMO, London (2007).

[17] INTERNATIONAL MARITIME ORGANIZATION, Revised Emergency Response Procedures for Ships Carrying Dangerous Goods (EmS Guide), 2018 Edition, MSC.1/Circ.1588, IMO, London (2018).

[18] INTERNATIONAL MARITIME ORGANIZATION, INTERNATIONAL LABOUR ORGANIZATION, WORLD HEALTH ORGANIZATION, Medical First Aid Guide for Use in Accidents Involving Dangerous Goods (MFAG), 1998 Edition, MSC/Circ.857, IMO, London (1998).

[19] INTERNATIONAL MARITIME ORGANIZATION, Lists of NAVAREA and METAREA Coordinators, COMSAR.1/Circ.58/Rev.1, IMO, London (2018).

[20] INTERNATIONAL MARITIME ORGANIZATION, International Code for the Safe Carriage of Packaged Irradiated Nuclear Fuel, Plutonium and High-Level Radioactive Wastes on Board Ships (INF Code), Resolution MSC.88(71), IMO, London (1999).

[21] INTERNATIONAL MARITIME ORGANIZATION, Guidelines for Developing Shipboard Emergency Plans for Ships Carrying Materials Subject to the INF Code, Resolution A.854(20), IMO, London (1997).

[22] Communication Dated 15 April 2014 Received from the Resident Representative of Norway to the Agency Regarding the Working Group on Best Practices for Voluntary and Confidential Government-to-Government Communications on the Transport of MOX Fuel, High Level Radioactive Waste and, as Appropriate, Irradiated Nuclear Fuel by Sea, INFCIRC/863, IAEA, Vienna (2014).

[23] INTERNATIONAL ATOMIC ENERGY AGENCY, Advisory Material for the IAEA Regulations for the Safe Transport of Radioactive Material (2018 Edition), IAEA Safety Standards Series No. SSG-26 (Rev. 1), IAEA, Vienna (in preparation).

[24] INTERNATIONAL ATOMIC ENERGY AGENCY, The Air Transport of Radioactive Material in Large Quantities or with High Activity, IAEA-TECDOC-702, IAEA, Vienna (1993).

[25] INTERNATIONAL ATOMIC ENERGY AGENCY, INTERNATIONAL CRIMINAL POLICE ORGANIZATION–INTERPOL, UNITED NATIONS INTERREGIONAL CRIME AND JUSTICE RESEARCH INSTITUTE, Radiological Crime Scene Management, IAEA Nuclear Security Series No. 22-G, IAEA, Vienna (2014).

[26] INTERNATIONAL ATOMIC ENERGY AGENCY, Developing a National Framework for Managing the Response to Nuclear Security Events, IAEA Nuclear Security Series No. 37-G, IAEA, Vienna (2019).

[27] INTERNATIONAL ATOMIC ENERGY AGENCY, Nuclear Forensics in Support of Investigations, IAEA Nuclear Security Series No. 2-G (Rev. 1), IAEA, Vienna (2015).

[28] INTERNATIONAL MARITIME ORGANIZATION, SOLAS: Consolidated Edition 2014: Consolidated Text of the International Convention for the Safety of Life at Sea, 1974, and Its Protocol of 1988: Articles, Annexes and Certificates, IMO, London (2014).

[29] INTERNATIONAL MARITIME ORGANIZATION, MARPOL: Consolidated Edition 2017: Articles, Protocols, Annexes and Unified Interpretations of the International Convention for the Prevention of Pollution from Ships, 1973, as Modified by the 1978 and 1997 Protocols, IMO, London (2017).

Annex I

REQUIREMENTS OF THE TRANSPORT REGULATIONS RELEVANT TO EMERGENCY ARRANGEMENTS

I–1. This annex summarizes the regulatory requirements in IAEA Safety Standards Series No. SSR-6 (Rev. 1), Regulations for the Safe Transport of Radioactive Material, 2018 Edition [I–1] (hereinafter referred to as the 'Transport Regulations'), that could influence the response to a transport emergency.

I–2. The transport of radioactive material is governed within States by national legislation. Since such transport may frequently involve transboundary operations, internationally agreed regulatory requirements have been developed. The Transport Regulations [I–1] are the basis for the safe transport of radioactive material in most States, by way of international modal and domestic transport regulations. The intention of the Transport Regulations [I–1] is that the packages will be designed, manufactured and maintained in such a way that, even in the event of an incident, the potential radiological impact will be acceptably small and, where fissile material is involved, criticality will be avoided.

I–3. The Transport Regulations [I–1] specify the basic design requirements to ensure safety during the transport of radioactive material. This is achieved by establishing requirements for the following (see para. 104 of the Transport Regulations [I–1]):

(a) Containment of the radioactive contents;
(b) Control of the external dose rate;
(c) Prevention of criticality;
(d) Prevention of damage caused by heat.

SHIPMENTS AND CONSIGNMENTS

I–4. In most shipments, radioactive material is transported in packages using normal handling procedures. However, some shipments have special characteristics that might affect the arrangements for emergency preparedness and response.

I–5. Some shipments are designated as 'exclusive use', as defined in the Transport Regulations [I–1]. These consignments are permitted to have a higher transport index (and hence higher external dose rates) and activity limits than would otherwise be allowed for the type of package being transported.

I–6. Some consignments may be transported under 'special arrangement' (i.e. in cases where it is impractical to ship the consignment in accordance with all the applicable requirements of the Transport Regulations [I–1]). The special arrangement provisions are required to compensate for not meeting all these requirements by providing an equivalent level of safety. Special precautions, administrative controls or operational controls are required, which may include emergency arrangements. Competent authority approval is required before transport; transboundary shipments also require multilateral approval prior to transport.

RADIOACTIVE MATERIAL

I–7. Special form radioactive material is either a non-dispersible solid radioactive material or a sealed capsule containing radioactive material. Special form radioactive material is required to withstand various tests, including impact, percussion, heat and bending tests, as applicable, and its design requires unilateral approval. Because of these requirements, special form radioactive material is regarded as unlikely to become dispersed and therefore unlikely to lead to a transport emergency.

I–8. Radioactive material classified as low specific activity (LSA)-I or as surface contaminated object (SCO)-I can be transported either packaged or unpackaged. An incident during the transport of either LSA-I or SCO-I (packaged or unpackaged) material is unlikely to lead to an emergency.

I–9. Radioactive material classified as SCO-III is a large solid object that, because of its size, cannot be transported in any of the types of package described in the Transport Regulations [I–1]. An example of SCO-III material is a disused steam generator or pressurizer from a nuclear power plant. An object classified as SCO-III is allowed to be shipped unpackaged, subject to certain conditions. This includes a requirement for emergency response and other special precautions associated with the shipment to be described in a transport plan. Competent authority approval (including approval of the transport plan) is required before transport, and international shipments require multilateral approval.

PACKAGES

I–10. The various types of package used for transporting radioactive material are described below. Depending on the type of package required, the graded approach used in the Transport Regulations specifies tests for the package design with respect to (i) routine conditions of transport (incident free), (ii) normal conditions of transport (minor mishaps) and (iii) accident conditions of transport.

Excepted packages

I–11. Excepted packages are permitted to contain only small quantities of radioactive material. The design requirements imposed on them are minimal, and excepted packages are exempt from most marking and labelling requirements. Typically, excepted packages are constructed of cardboard or fibreboard. Examples are packages that contain radioactive consumer products, radiopharmaceuticals and very low activity radioactive sources used for testing instruments. Empty packaging that is internally contaminated may also be transported as an excepted package. An incident during the transport of excepted packages is unlikely to lead to an emergency. However, the packages still need to be handled with caution after an incident, as contamination might be present.

Industrial packages

I–12. Industrial packages are permitted to contain material that has a low activity per unit mass (known as LSA material) or non-radioactive items that have low levels of surface contamination, known as SCOs, relating to classifications SCO-I or SCO-II.

I–13. The quantity of LSA, SCO-I or SCO-II material allowed in a single industrial package is restricted so that the external dose rate at 3 m from the unshielded material does not exceed 10 mSv/h. In an emergency during the transport of this type of material, the radiological consequences will therefore be limited.

I–14. Three types of industrial package are defined in the Transport Regulations [I–1]: Type IP-1, Type IP-2 and Type IP-3. The testing requirements and maximum activity limits increase from IP-1 to IP-3 (see paras 623–630 of the Transport Regulations [I–1]). The type of industrial package that is permitted depends on the characteristics of the LSA material or the SCO to be transported.

I–15.   Although the specific activity of LSA material and the contamination of SCOs is generally low, the total activity in a consignment could, in some cases, be significant. Some examples of LSA material and SCOs are as follows:

(a)   LSA-I: Can be solid or liquid. LSA-I material typically includes ores, unirradiated uranium and thorium, mill tailings, and contaminated soil and debris with low activity concentrations. The material usually has uniform activity distribution.
(b)   LSA-II: Can be solid or liquid. LSA-II material typically includes reactor process wastes, filter sludges, absorbed liquids and resins, activated equipment, laboratory wastes and decommissioning wastes. This material often has a less uniform activity distribution than LSA-I; i.e. localized higher activity concentrations might be present and more stringent packaging requirements are imposed.
(c)   LSA-III: Solid material (excluding powders) only. LSA-III material typically includes solidified liquids, resins, cartridge filters and irradiated material. This material is required to be essentially uniformly distributed in a solid compact binding agent. Radioactive material may also be distributed throughout a single solid object or a collection of solid objects within the packaging. This material is allowed to have higher specific activities; therefore, more stringent packaging requirements and restrictions on the material characteristics are imposed.
(d)   SCO-I, SCO-II and SCO-III: These groups cover solid objects that are not radioactive themselves but which have contamination on their surfaces. SCO-II allows for higher contamination levels than SCO-I. Examples would be decommissioning waste such as contaminated piping, tools, valves, pumps and other hardware. Material classified as SCO-III is a large solid object that, because of its size, cannot be transported in a package (see para. I–9).

I–16.   All industrial packages are required to meet general packaging requirements. Type IP-2 and Type IP-3 are required to withstand the normal conditions of transport (i.e. including minor mishaps) without a loss or dispersal of their contents or a loss of integrity of any radiation shielding. Typical examples of industrial packages are steel drums and plastic or metal bulk containers and tanks.

Type A packages

I–17.   Type A packages are permitted to contain limited quantities of radioactive material. These activity limits are based on the maximum acceptable radiological consequences following a failure under accident conditions of transport. Activity limits are specified in the Transport Regulations [I–1] for each radionuclide. Separate limits are specified for special form radioactive material (see para. I–7) and radioactive material other than special form radioactive material. These limits are known as the A_1 and A_2 values, respectively.

I–18. Type A packages are required to withstand normal conditions of transport without a loss or dispersal of their contents or a loss of adequate shielding integrity. They are not specifically designed to withstand accident conditions, except when containing liquids or gases. They range from wood or fibreboard constructions with glass, plastic or metal inner containers to metal drums or lead filled steel containers. Examples of material transported in Type A packages include radiopharmaceuticals, radionuclides for industrial applications, and some types of radioactive waste.

Type B(U) and Type B(M) packages

I–19. Type B(U) and Type B(M) packages are permitted to contain radioactive material in quantities greater than those allowed in Type A packages. Type B(U) and Type B(M) packages are required to be designed to withstand both normal conditions and accident conditions of transport (which are simulated by drop, puncture, crush, thermal and immersion tests). Type B(U) and Type B(M) packages range from small containers of a few kilograms (e.g. containing industrial radioactive sources) to large packages up to about 100 t (e.g. containing spent fuel from nuclear power plants). Typically, Type B(U) and Type B(M) packages are of steel construction and incorporate substantial radiation shielding. The Transport Regulations [I–1] require Type B(U) and Type B(M) package designs to be approved by the relevant competent authorities.

Type C packages

I–20. Type C packages are designed to transport radioactive material with high levels of activity by air. These packages are designed to withstand the drop, puncture, thermal and immersion tests for Type B(U) and Type B(M) packages; in addition, they are designed to withstand more severe tests intended to simulate the conditions that result from a severe aircraft accident. Type C package designs are subject to approval by the competent authority.

Packages containing uranium hexafluoride

I–21. Uranium hexafluoride in quantities of 0.1 kg or more is required to be packaged and transported in accordance with the provisions of Ref. [I–2] as well as the relevant requirements of the Transport Regulations [I–1]. Designs for packages that will contain 0.1 kg or more of uranium hexafluoride are subject to approval by the competent authority.

I–22. An emergency involving uranium hexafluoride primarily presents a chemical hazard.

Packages containing fissile material

I–23. Fissile material for the purposes of transport is a material containing any of the fissile nuclides, namely ^{233}U, ^{235}U, ^{239}Pu and ^{241}Pu, with some exceptions listed in the Transport Regulations [I–1]. Fissile material is capable of undergoing a self-sustaining nuclear chain reaction under certain conditions, resulting in the release of radiation and heat.

I–24. Packages containing fissile material may be industrial packages or Type A, Type B(U), Type B(M) or Type C packages. The design of any package intended to contain fissile material is subject to the approval of the competent authority, with some exceptions (see paras 417, 674, 675 of the Transport Regulations [I–1]).

I–25. The Transport Regulations [I–1] include specific requirements for packages containing fissile material, which are intended to ensure criticality safety through the following provisions:

(a) Limiting the quantity and geometric configuration of the fissile material;
(b) Imposing strict package design features to ensure that criticality safety is maintained under the tests for accident conditions;
(c) Controlling the number of packages that are permitted to be carried on a single conveyance or that are permitted to be stowed together during transport and in-transit storage.

The Transport Regulations [I–1] contain some exceptions to these requirements for packages containing fissile material, for example if the ^{235}U concentration does not exceed 1%, or if the package contains only limited quantities of fissile material. These are known as 'fissile excepted' packages. In these cases, the other relevant requirements of the Transport Regulations [I–1] that relate to the radioactive nature of the contents are still applicable.

DOSE RATES AND CATEGORIES

I–26.   The dose rate limit for excepted packages is 5 µSv/h at any point on the external surface of the package.

I–27.   The dose rate under routine conditions of transport is not allowed to exceed 2 mSv/h at any point on the external surface of the vehicle or freight container, and 0.1 mSv/h at 2 m from the external surface of the vehicle or freight container.[1]

I–28.   Maximum dose rates form the basis of the labelling categories for packages, overpacks and freight containers (see paras I–31 to I–35), as summarized in Table I–1. These label categories provide information to assist in ensuring adequate radiation protection during handling, stowage and storage of the packages. The categorization of packages can also assist emergency workers in understanding the level of risk posed by undamaged packages in an emergency.

TABLE I–1. MAXIMUM DOSE RATES FOR EACH TYPE OF PACKAGE LABEL

Category of label	Conditions of transport		Maximum dose rate at any point on the external surface of the package(mSv/h)	Transport index
	Under exclusive use	Not under exclusive use		
I-WHITE		X	Not more than 0.005	0
II-YELLOW		X	More than 0.005 but not more than 0.5	More than 0 but not more than 1
III-YELLOW		X	More than 0.5 but not more than 2	More than 1 but not more than 10
III-YELLOW	X		More than 2 but not more than 10	More than 10

---

[1]   These dose rate limits do not apply to consignments transported under exclusive use or special arrangement.

I–29.   For Type IP-2, Type IP-3, Type A, Type B(U), Type B(M) and Type C package designs, the Transport Regulations [I–1] require that the maximum dose rate on the external surface not increase by more than 20% when such packages are tested to withstand the normal conditions of transport. For Type B(U), Type B(M) and Type C package designs, it is required that the dose rate not exceed 10 mSv/h at 1 m from the package surface when such packages are tested to withstand the accident conditions of transport. These requirements help ensure the protection of the public and emergency workers during an emergency involving these types of package.

MARKING OF PACKAGES

I–30.   For all packages — other than excepted packages transported by post (which are permitted to carry only very small quantities of radioactive material) — the UN number is required to be legibly and durably marked on the outside of the packaging. The package is also required to be marked with an identification of the consignor or the consignee, or both. Each package of a gross mass exceeding 50 kg is required to have its permissible gross mass legibly and durably marked on the outside of the package. In addition, packages are required to be legibly and durably marked with the package type on the outside of the packaging. The marking requirements in the Transport Regulations [I–1] for different types of package are summarized in Table I–2.

TABLE I–2. MARKING REQUIREMENTS FOR PACKAGES
CONTAINING RADIOACTIVE MATERIAL

Marking	Package type							
	Excepted	Type IP-1	Type IP-2	Type IP-3	Type A	Type B(U)	Type B(M)	Type C
Consignor or consignee identification, or both	X	X	X	X	X	X	X	X
UN number	X	X	X	X	X	X	X	X
Proper shipping name		X	X	X	X	X	X	X

TABLE I–2. MARKING REQUIREMENTS FOR PACKAGES
CONTAINING RADIOACTIVE MATERIAL (cont.)

Marking	Package type							
	Excepted	Type IP-1	Type IP-2	Type IP-3	Type A	Type B(U)	Type B(M)	Type C
For package mass greater than 50 kg, permissible gross mass	X	X	X	X	X	X	X	X
Type IP-1, IP-2, IP-3, A, as appropriate		X	X	X	X			
VRI[a] code of country of design origin and name of manufacturer			X	X	X			
Competent authority identification for design	X[b]	X[b]	X[b]	X[b]	X	X	X	X
Serial No.	X[b]	X[b]	X[b]	X[b]	X	X	X	X
Type B(U), B(M), C, as appropriate						X	X	X
Trefoil symbol						X	X	X

[a] Vehicle registration identification.
[b] The requirement applies only if the package contains fissile material or 0.1 kg or more of uranium hexafluoride.

LABELLING OF PACKAGES

I–31. Packages (other than excepted packages), freight containers and overpacks containing radioactive material are required to bear labels indicating their category (see para. I–28): I-WHITE, II-YELLOW or III-YELLOW. The I-WHITE label indicates very low dose rates outside a package, whereas II-YELLOW and (especially) III-YELLOW labels indicate higher dose rates (see Table I–1) that might be significant in terms of the response to an emergency. In addition to these labels, packages containing fissile material — if not excepted from the fissile material requirements — are required to bear a label indicating that they contain fissile material. These labels are shown in Fig. I–1.

I–32. The labels provide information on the external radiation hazards associated with undamaged packages. This information is used to control the manner in which packages of radioactive material are handled and stowed during transport and stored during in-transit storage. However, the same information can also assist with the response in the event of a transport emergency.

FIG. I–1. Labels used on packages containing radioactive material and fissile material [I–1].

I–33.   The labels are also required to display the radionuclides and the total activity in the package. For categories II-YELLOW and III-YELLOW, the labels are required to include the transport index (TI). The TI is a number used in the control of radiation exposure and is an indicator of the dose rate at 1 m from the external surface of the package.

I–34.   Packages containing fissile material are also required to bear criticality safety labels, also shown in Fig. I–1, which display the criticality safety index (CSI) as stated in the certificate of approval issued by the competent authority. The CSI is a number that provides information to assist in the control of criticality.

I–35.   Packages containing radioactive material that has other dangerous properties are required to also bear appropriate labels in compliance with the relevant transport regulations for dangerous goods.

PLACARDING

I–36.   Rail and road vehicles carrying any labelled packages, large freight containers containing packages other than excepted packages, tanks containing radioactive material, and certain consignments of LSA-I or SCO-I material in large freight containers or tanks are required to bear placards indicating the presence of radioactive material, as shown in Fig. I–2. The UN number for the consignment is also required to be displayed in certain cases.

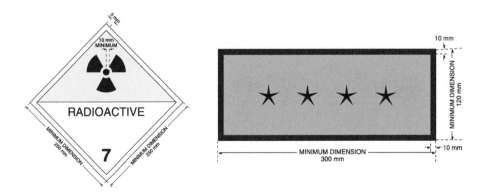

FIG. I–2. Placards used on vehicles, tanks and freight containers carrying radioactive material. The symbol "****" denotes the space in which the appropriate UN number for radioactive material is displayed [I–1].

TRANSPORT DOCUMENTS

I–37. Consignors are required to provide transport documents for each consignment. These documents include the particulars of the consignment, the consignor's certification or declaration, and information for carriers, including the emergency arrangements appropriate for the consignment (this does not apply to excepted packages). These transport documents are required to be provided by the consignor to the carrier. This information can assist those responding to an emergency with identifying the contents of the consignment and help ensure proper response to the emergency. In some cases, the information might not be immediately available at the emergency site area (e.g. if the documents have been destroyed by the initiating event). If so, this information will need to be sought from the consignor.

REFERENCES TO ANNEX I

[I–1] INTERNATIONAL ATOMIC ENERGY AGENCY, Regulations for the Safe Transport of Radioactive Material, 2018 Edition, IAEA Safety Standards Series No. SSR-6 (Rev. 1), IAEA, Vienna (2018).

[I–2] INTERNATIONAL ORGANIZATION FOR STANDARDIZATION, Nuclear Energy — Packagings for the Transport of Uranium Hexafluoride (UF$_6$), ISO 7195:2020, ISO, Geneva (2020).

Annex II

EXAMPLE EVENT NOTIFICATION FORM

II–1.   This annex contains an example of an event notification form. Based on the national hazard assessment, event notification forms can be developed to meet the needs of each State and response organization.

II–2.   Event notification forms are intended to be completed as information on a transport emergency becomes available. Initial information provided by emergency services or the carrier is provided to the notification point as soon as a visual inspection of the site has been undertaken. The notification point can then relay this information to other response organizations, and they can use the information to determine whether additional resources are needed for an effective emergency response. The information gathered can also be used in instructions, warnings and information for the public.

II–3.   As the response efforts continue and more information becomes available, the form is updated.

II–4.   Part 1 of the form focuses on gathering information while initial response actions are being taken. The information is based on observables and indicators, not on measurements. Depending on the nature of the emergency, the information could be provided by the carrier or by first responders.

II–5.   Part 2 of the form focuses on allowing the radiological assessor to prepare to deploy to the site area and to provide advice and assessments. Depending on the nature of the emergency, the information may be provided by a combination of sources, including emergency workers at the site and offices of the carrier or consignor.

	Event Notification Form	
	PART 1	
	*Note: Do not delay emergency response actions or additional notifications to complete this form. Gather the information that is readily available.*	
1.1	Name and contact information of person or agency reporting the event:	
1.2	Date and time of the event: *Specify time zone*	
1.3	Exact location of the emergency: *Address or GPS coordinates*	
1.4	Information on the conveyance(s): *Registration number, International Maritime Organization number, flight number*	
1.5	Description of the event: *Collision, sinking, etc.*	

1.6	Key observable conditions:
	 — Description of package(s).
	  Drums, cardboard boxes, etc.
	  Number, if known.
	 — Fire?
	  Duration? Extinguished?
	 — Visible damage to package?
	  Describe.	

	— Suspected release of package contents? *Leak, spill, venting, etc.* — Status of the conveyance(s) and package(s). *Overturned, sunken, etc.* — Presence of other dangerous goods?	
1.7	Description of initial response actions: *Number of victims, rescues? First aid? Cordoned off area established? Note: Confirm that there is no delay in transport of injured victims due to possible contamination.*	
1.8	Transport document information: — Number of package(s) — Radionuclide(s) and activities — Carrier information — Consignor information — Consignee information — UN number — Chemical and physical forms	
1.9	Marking and labelling information: — Label category (I-WHITE, II-YELLOW, III-YELLOW) — Contents of labels — UN number — Radionuclide(s) and activities — Transport index — Criticality safety index (if applicable)	
1.10	Weather conditions: *Rain, storms, strong wind, etc.*	
1.11	Photographs and/or sketch of the site area:	
1.12	Photographs and/or sketch of the package(s), including labels and markings:	

PART 2	
Note: Continue to gather information from Part 1 if this was not previously available.	
2.1	Detailed description of initiating event: *Drop height, collision speed, fire duration, etc.*	
2.2	Status of protective actions and other response actions:	
2.3	Description of transport modality: *Modes, route, locations, etc.*	
2.4	Additional information on: — Package(s) — Freight container(s) — Conveyance	
2.5	Measurement results, if available: — Dose rates — Contamination surveys	
2.6	Package type(s) and design certificate:	
2.7	Other hazards present at the site: *Severe weather, conventional hazards, etc.*	
2.8	Accessibility of the site:	
2.9	Additional meteorological data:	

2.10	Logistical support available at the site:	
2.11	Description of surrounding area: — Population — Essential infrastructure — Agriculture — Drinking water supply — Protected or restricted areas	

Annex III

TEMPLATE FOR THE CARRIER OR CONSIGNOR
EMERGENCY RESPONSE PLAN

III–1.   This annex contains a template for the emergency response plan prepared by either the consignor or the carrier of consignments of radioactive material, in accordance with paras 304 and 305 of the Transport Regulations (IAEA Safety Standards Series No. SSR-6 (Rev. 1), Regulations for the Safe Transport of Radioactive Material, 2018 Edition [III–1]).

TITLE (COVER) PAGE

III–2.   The title (cover) page contains the title of the plan, the approval date, the version number and relevant signatures. The signatures could include the heads of all the participating organizations.

INTRODUCTION

III–3.   This section describes the objectives and contents of the plan. This section also states the scope of the plan and what phases of the emergency it covers. It also describes the relevant regulatory or legal framework.

III–4.   This section also lists the individuals responsible for implementation and maintenance of the plan.

OVERVIEW OF SHIPMENTS

III–5.   This section provides a general description of the different types of package that will be transported, along with how they are to be handled. Documents that could provide more information are also referred to, along with where to find these documents.

INTERNAL ORGANIZATION OF THE RESPONSE

III–6. This section presents the consignor or carrier provisions for managing an emergency. These provisions need to be consistent with any emergency plans prepared by national, regional and local authorities.

III–7. The following key points need to be addressed in this section:

(a) The organizational approach to detecting a possible event that could lead to an emergency, and the dissemination of the subsequent alert;
(b) The organizational approach to the response following the alert, both for the initial phase and the longer term;
(c) The organizational approach to a long term emergency;
(d) The organizational approach to the termination of an emergency.

III–8. The roles and responsibilities of each party with a role in the emergency response are presented in this section, which also specifies the measures taken to guarantee the availability of sufficient personnel and resources for an effective response.

III–9. This section also describes the locations of response organizations and persons, the scope of their responsibilities for decision making (including the extent of external communication) and the interactions that take place between different parties.

III–10. The interactions with national, regional and local authorities and the procedures involved are also specified, including flow charts and organizational diagrams, as appropriate.

PROCEDURES FOR TRIGGERING THE PLAN AND MAKING NOTIFICATIONS

III–11. This section describes the means for detecting an incident during the transport of radioactive material that could lead to an emergency. It also describes the criteria for activating the emergency response plan and the procedures for alerting response organizations and public authorities when an emergency occurs.

III–12. Notification points and notification procedures are also described in this section (including an event notification form, such as that provided in Annex II).

EMERGENCY RESPONSE

Response personnel

III–13. This section describes the capacity to deploy personnel with the necessary skills and experience for the emergency response. It states which parties are likely to be involved, their training and qualifications, and a time frame for their deployment.

Emergency scenarios

III–14. This section describes the postulated emergency scenarios considered in the development of the consignor or carrier's emergency response plan and associated arrangements.

Resources available for deployment to the site area

III–15. This section lists any equipment necessary to respond to the emergency scenarios considered in the plan. This includes the equipment needed during each phase of the emergency, and the time and resources needed to make this equipment available.

Provisions for the emergency response

III–16. This section specifies the steps to be taken to respond to an emergency.

Interim location(s) for damaged packages

III–17. This section identifies the characteristics of interim locations where damaged packages could be moved while maintaining an adequate level of safety. This section describes any existing agreements with such locations, along with the steps necessary to gain authorization for the movement of damaged packages.

Termination of an emergency

III–18. This section describes the provisions for termination of an emergency, including any measures for transition to a planned exposure situation or an existing exposure situation, if such a transition is considered necessary.

EMERGENCY MANAGEMENT TOOLS

III–19. This section describes the operational tools that are available to help manage the emergency; examples are given in paras III–20 to III–24.

Decision aiding tools

III–20. Decision aiding tools could include practical tools, such as logic diagrams, to help direct the response actions that are taken.

Response procedure

III–21. The response procedure outlines each step in the emergency plan, for each party involved in the emergency response, in chronological order. It includes details of the conditions for using the procedure, the expected results and the conditions for ending use of the procedure.

Standard messages

III–22. The provision of standard messages encourages a standard approach for transmitting messages and for the information to be provided, such as date, time, sender details, reference, event details and details of the response actions taken.

External communication

III–23. This section describes the arrangements for external communications with the public, news media, and national, regional and local authorities.

Recording and archiving of communications

III–24. This section describes how the various communications are logged during management of the emergency, and how they are archived and made available.

MAINTAINING OPERATIONAL READINESS

III–25. This section describes how operational readiness to respond to an emergency will be maintained.

Training of personnel

III–26.   This section includes details of the training of all personnel described in the plan and the provisions to ensure that a sufficient number of qualified and trained personnel are always available to implement the plan.

Exercises

III–27.   This section describes the exercises needed to test the plan's adequacy and the intervals at which they will take place. The frequency and scope of exercises testing different areas of the plan are described, as is the level of involvement of response organizations and other parties.

Experience feedback

III–28.   This section describes how learning from exercises, actual emergencies and other sources of information is taken into account in the plan.

Renewal of partnerships

III–29.   The procedures for the renewal of any partnerships or agreements are specified in this section.

Quality assurance

III–30.   This section describes how the quality of the plan is maintained and includes provisions for the management of documentation relating to quality assurance.

ANNEX I TO THE CARRIER OR CONSIGNOR EMERGENCY RESPONSE PLAN

III–31.   Contact information of national emergency contact points is provided in this annex.

ANNEX II TO THE CARRIER OR CONSIGNOR EMERGENCY RESPONSE PLAN

III–32.   The event notification form is provided in this annex (see Annex II of this Safety Guide).

REFERENCE TO ANNEX III

[III–1]   INTERNATIONAL ATOMIC ENERGY AGENCY, Regulations for the Safe Transport of Radioactive Material, 2018 Edition, IAEA Safety Standards Series No. SSR-6 (Rev. 1), IAEA, Vienna (2018).

Annex IV

POSTULATED EVENTS AND POTENTIAL CONSEQUENCES FOR THE HAZARD ASSESSMENT

IV–1. The following events are hypothetical emergency scenarios based on a combination of events that have occurred and postulated plausible events. The radiological consequences can be considered independent of their initiator. The examples presented in this annex are intended to be representative so that emergency planners and transport safety experts can develop the emergency planning basis and determine the associated emergency arrangements. The radioactive material and modes of transport in these scenarios are not based on any study of probability or likelihood. Each State needs to conduct its own hazard assessment based on the modalities of the transport within its territory.

IV–2. On the basis of the hazard assessment developed for different accident scenarios, an appropriate concept of operations for each scenario can be developed. When an emergency involving a shipment of radioactive material occurs, the applicable response actions are then implemented. The response actions and the equipment required to implement them effectively need to be made available by the national emergency response authority to its network of regional and local response units, including first responders.

SCENARIO 1: A HIGH ENERGY COLLISION, PLUS FIRE, INVOLVING A TYPE B(U) PACKAGE BEING TRANSPORTED BY ROAD

IV–3. A road vehicle carrying international cargo is involved in a high energy collision, followed by a fire lasting about one hour. The vehicle is badly damaged, and the driver and the driver's assistant are injured. The vehicle placards are obscured by the fire.

IV–4. The first responders arrive at the site area, rescue injured persons, extinguish the fire and only then observe that the vehicle is carrying radioactive material. Looking at the marking and labelling on the package, they identify a Type B(U) package containing ^{137}Cs sources. The first responders notify the emergency response centre and establish and cordon off the site area. Radiological assessors are mobilized, and the team of assessors immediately proceeds to the site area.

IV–5. On reaching the site area, the team of radiological assessors survey the site and the first responders and confirm that there is no contamination. The team also visually assesses the package and concludes that it appears to be intact but needs to be assessed further to ensure that all the safety functions are intact. The team confirms that there is no contamination on the package surface, confirms that the dose rate measurements are consistent with the information in the transport documents (obtained from the consignor), retightens loose closures on the package and forwards the package to a secure interim location for further assessment. The site area is reopened to the public as soon as the damaged vehicles have been removed, approximately ten hours after the incident occurred.

Potential consequences

IV–6. The vehicle crew and the response personnel could have received significant radiation exposure if the package had been damaged.

IV–7. If the radioactive material was in dispersible form, radioactive contamination of the environment might have occurred.

IV–8. Before first responders arrived at the site area, some individuals in the immediate vicinity of the site could have been exposed to radiation.

IV–9. If the environment had been contaminated, members of the public who were present around the site area could have received internal radiation exposure.

IV–10. When response actions were being taken, the response personnel might have been exposed to radiation, depending on their distance from the package and the length of time they remained at that distance from the package.

IV–11. The radiation dose received could be estimated and, where possible, verified from measurements taken using appropriate instrumentation.

SCENARIO 2: DERAILMENT OF A CONSIGNMENT OF URANIUM ORE CONCENTRATE BEING TRANSPORTED BY RAIL

IV–12. A railway wagon carrying uranium ore concentrate is derailed, and this results in injuries to the train crew and in the railway track being blocked. The derailed wagon is carrying 50 industrial packages of Type IP-1, each of which is a 200 L drum containing low specific activity (LSA)-I material. Twelve drums are thrown out of the wagon by the derailment; the other 38 remain in the wagon.

The 12 ejected drums land 1–10 m from the vehicle and suffer different levels of damage. Some have visible holes and puncture marks. The accident occurs in a remote location, and the weather is wet.

IV–13.   The first responders from the nearest town reach the site area. From the placards on the rail vehicle, they identify the radioactive contents in the packages. They establish and cordon off the site area and notify the appropriate authorities, who arrange for radiological assessors and representatives of the consignor to attend the scene. The train crew who were injured in the incident are rescued and transported to a hospital for treatment.

IV–14.   On reaching the site area, the radiological assessors confirm that there is no contamination on the first responders. However, there is some contamination near the railway track. The site area is kept cordoned off while the consignor deploys resources to clean up the site. The spilled ore concentrate is recovered and placed in new drums. The undamaged drums are surveyed and transferred to a new rail vehicle. The railway track is subsequently reopened when all the damaged packages have been removed, approximately one day after the incident occurred. Meanwhile, radiological assessors proceed to the hospital, and it is confirmed that there is no contamination of the injured train crew, the ambulance or its crew, or the hospital.

Potential consequences

IV–15.   In this incident, there was radioactive contamination near the railway track where the drums fell out of the wagon. There could also have been contamination on the wagon itself if any of the drums on the wagon suffered an impact during the incident.

IV–16.   Contamination can result in internal exposure; however, with LSA-I material, any such exposure is likely to be low over an exposure period of approximately one day. Wet weather would have helped in this respect by preventing resuspension of dust.

IV–17.   The location of the site area could impede communication between personnel at the site and the local and regional emergency response units, as well as delay the arrival of first responders, the radiological assessors and the consignor. However, this location means that public exposure would not occur, and the site area could be easily cordoned off and maintained during the emergency response.

IV–18.   The wet weather could interfere with the response actions and result in the spread of contamination through runoff of contaminated surface water.

IV–19.   Any radioactive waste arising from the decontamination of the site area would need to be collected, assayed and sent for disposal via an appropriate disposal route.

SCENARIO 3: ROAD TRANSPORT ACCIDENT INVOLVING TYPE IP-2 PACKAGES

IV–20.   A truck carrying low level radioactive waste in industrial packages of Type IP-2 accidentally leaves the road and crashes down a 10 m embankment, plunging into the stream below. The packages are ruptured, and the radioactive contents are spread on the embankment. Some of the contents remain in the truck, which is partially submerged in 1 m of water in the stream. The first responders rescue the driver, notify the emergency response centre and establish and cordon off the site area.

IV–21.   Radiological assessors arrive at the site area within a few hours. They set up temporary dykes in the stream. They survey the embankment and take water samples. They observe that radioactive contamination has spread over approximately 500 m^2 of land. The water samples taken a few metres downstream show very slightly elevated levels of radioactivity. The public are instructed not to swim in the stream, use the water, or fish until further notice. Contaminated surface soil up to a depth of 10 cm is removed, placed in boxes and sent for safe disposal. The area is closed to the public for four days while the decontamination of the site area is completed. Thereafter, the area is declared safe for public use; all restrictions on using the stream are withdrawn.

Potential consequences

IV–22.   The truck driver's skin and clothes have minor contamination from initial response actions in the water near the packages. External exposure is limited because of the nature of the radioactive material.

IV–23.   Air samples and water samples are collected and analysed to confirm that there is no long term residual contamination.

SCENARIO 4: ROAD TRANSPORT ACCIDENT INVOLVING EXCEPTED PACKAGES AND TYPE A PACKAGES CONTAINING RADIOPHARMACEUTICALS

IV–24. A delivery van carrying a consignment of radiopharmaceuticals is involved in a road accident. The vehicle is carrying a total of 82 Type A packages and excepted packages originating from five consignors for delivery to a number of medical institutions. The severity of the impact causes all the cargo to be ejected and dispersed on both sides of the road over a distance of about 200 m. Thirty packages are damaged. Two of these Type A packages — one containing ^{67}Ga (200 MBq) and the other ^{131}I (40 MBq) — suffer a loss of containment. Vials containing radioactive material escape from their shielding and are subsequently broken.

IV–25. A crew member contacts the local police and the relevant emergency management agency. Within 15 minutes, police officials reach the site area, followed by the local fire department. A representative of the local civil defence department arrives at the site area, equipped with a radiation monitor. A superficial survey confirms elevated levels of radiation at the site area. The police cordon off the area and wait for radiological assistance.

IV–26. The emergency management agency notifies the appropriate response centre about the incident. A radiation protection team reaches the site area within two hours of the occurrence of the accident. Using the information in the transport documents, the team prepares an inventory of the sources involved in the event and conducts an extensive survey of the area with a suitable monitoring instrument. The emergency vehicles at the site area, the civil defence personnel, the police officers and the damaged van are also surveyed: no contamination is detected. The survey of the site area indicates localized contamination from the leaking vials but concludes that there is no public health hazard.

IV–27. Under the guidance of the radiological assessors, small pieces of contaminated debris and packing material are collected in plastic bags, then placed in cardboard boxes and sent for safe disposal together with the damaged packages. In the area where the ^{131}I source was broken, approximately 0.1 m^3 of topsoil is removed, placed in boxes and sent for safe disposal. A thorough and systematic survey of the area is then carried out. Normal background dose rates are measured. Sixteen hours after the accident occurred, and after being thoroughly washed, the highway is reopened for public use.

Potential consequences

IV–28.   The potential radiological hazards arising from an accident involving several Type A packages could be potentially significant. In this case, damage to Type A packages containing radiopharmaceuticals (unsealed sources), accompanied by the release of the contents, could result in both internal and external exposure. Even if there was only damage to the shielding and not the containment, external exposure could still occur.

IV–29.   The vehicle crew, the bystanders and the response personnel might have received some radiation exposure. Similarly, the decontamination of the area and the collection of radioactive waste would contribute to the exposures received by the emergency workers.

IV–30.   Spread of contamination could occur owing to wind and the movement of vehicles on the road. The latter can be minimized by stopping vehicular movement on the road until the emergency is terminated.

IV–31.   A problem that may be encountered in an accident of this type is the possible lack of information concerning the exact composition of the consignment. It is typical for a carrier to make several deliveries and pickups during a particular assignment. The original integrated bill of lading, therefore, may not correctly indicate the exact contents at various stages of the journey (e.g. after the first delivery or the second pickup has been made).

SCENARIO 5: INCIDENT INVOLVING AIR TRANSPORT OF ^{192}Ir PELLETS IN A PACKAGE WITH REDUCED SHIELDING

IV–32.   Iridium-192 pellets within a lead shielded Type B package are being shipped from State A to State C by air via State B, and then on to their final destination in State C by road. En route by road in State C, the driver's personal dosimeter sounds an alarm. The driver stops the vehicle, moves 30 m away, and calls the first responders, in accordance with the emergency instructions.

IV–33.   The first responders arrive at the site area and, using the information given to them by the driver, cordon off an area of 30 m radius. In accordance with pre-established arrangements between the consignor and consignee, the consignee sends radiological assessors to the site area.

IV–34.   Dose rates at one part of the 30 m cordoned off area are found to be 5 mSv/h; hence, the cordon is expanded by the first responders to a radius of 100 m, where a lower dose rate of 100 µSv/h is measured. Dose rate variations along the cordon indicate that one side of the package has lost its shielding function, for unknown reasons. The consignee is able to apply temporary additional shielding to the package and moves it to the final destination. The roadway is reopened to the public six hours after the initial actions by the driver.

IV–35.   The competent authorities in all three States are notified, and the personnel who handled the shipment are identified. Blood samples are taken for biodosimetry, and a total of four employees from all three States are shown to have received individual effective doses of approximately 100 mSv.

Potential consequences

IV–36.   The reason for the reduction in shielding integrity is not known. Such an event could have been initiated by an operational error or by equipment being in poor condition owing to lack of maintenance.

IV–37.   An incident like this involves external exposure to package handlers, vehicle crew, the persons responding to the accident and bystanders. The dose rates at different distances from the package can be used to estimate the doses received by persons. The driver would have received external exposure while at the wheel, while attending to the vehicle when it was stationary and while loading the package into the vehicle.

IV–38.   During the period when the cordon was established at only 30 m, persons could have been exposed to increased radiation levels. The doses received by persons present between 30 and 100 m need to be determined.

IV–39.   The possible doses received by persons at the airports in States A, B and C, where the consignment was handled, also need to be estimated. However, without information on when and where the impairment of shielding integrity occurred, the estimated exposures of these persons are subject to large uncertainties.

IV–40.   The exposure of some persons can be directly assessed from the results of personal dosimetry, where available; this includes the driver (if a dosimeter was worn) and the persons who installed the additional shielding during the emergency response.

SCENARIO 6: COLLISION OF A ROAD VEHICLE CARRYING A URANIUM HEXAFLUORIDE PACKAGE, FOLLOWED BY FIRE

IV–41.   A truck carrying a 48Y cylinder containing 12 t of natural uranium hexafluoride (UF_6) is involved in a collision with a mobile tank containing liquid hydrocarbon fuel. The collision results in a fire engulfing the 48Y cylinder. The truck driver, who is only slightly hurt, notifies the national emergency contact points for radiological safety and nuclear security related events, as well as the consignor. The public authority immediately alerts the local fire brigade and the other organizations identified in the local emergency plan, including radiological assessors and an expert in the chemical toxicity of hydrogen fluoride (HF). This plan includes a cordoned off area of 100 m radius and sheltering downwind up to at least 1000 m from the package. The fire brigade intervenes upwind to fight the fire.

IV–42.   Approximately one hour after the initial collision, the cylinder ruptures and disperses an unknown quantity of UF_6 as liquid and vapour in the downwind direction. The UF_6 reacts with the moisture in the air, generating HF and uranyl fluoride (UO_2F_2).

IV–43.   After the rupture of the cylinder, the fire brigade stops using water on the fire and spreads foam instead. In addition, water is sprayed downwind to flush down any remaining emission of HF and UO_2F_2.

IV–44.   Approximately 90 minutes after the initial collision, the fire is extinguished.

IV–45.   The radiological assessors who arrived at the site area take air and ground samples outside the cordoned off area, which indicate contamination in the downwind direction up to a distance of several kilometres. The cordoned off area is then extended accordingly.

IV–46.   The people and emergency workers present in the contaminated area during the passage of the HF and UO_2F_2 plume are sent to hospital to be checked for possible chemical and radiation exposure.

IV–47.   The consignor makes arrangements for recovery (e.g. preparation of a rescue container) to secure the package before moving it to a safe location.

IV–48. UO_2F_2 deposits within and outside the site area are removed and placed in drums, and the vehicle wrecks are removed from the site area three days after the incident. The wrecks are sent to a cordoned off area in a nearby scrapyard for decontamination. The damaged cylinder is moved from the site area to a safe location. Cleaning and decontamination of the road is then performed, and the road is reopened.

Potential consequences

IV–49.   The fire could release a significant quantity (between 8 and 12 t) of UF_6, which then converts to HF and UO_2F_2. Persons present downwind of the cylinder could have incurred an intake of these chemicals, as could those who were engaged in firefighting, in decontamination activities and in handling the damaged cylinder and vehicles. Inhalation of these chemicals can represent a significant hazard. In contrast, the radiological hazards would be a subsidiary (but not negligible) hazard.

IV–50.   Rupture of the cylinder could have occurred before the effective sheltering of all persons within 1 km downwind. Measured values of the concentrations of UF_6, HF and UO_2F_2 in the environment, and the duration of exposure of persons to these compounds, would be useful in determining the intake; however, it is often unreasonable to expect such measurements one hour after the accident. Air concentrations can also be estimated by calculation, based on the wind speed and atmospheric stability class, using realistic assumptions for the plume heights and release rates. To estimate possible intakes, the use of respiratory protective equipment and sheltering also needs to be taken into account.

SCENARIO 7: SINKING OF A CARGO VESSEL CONTAINING ^{137}Cs RADIOACTIVE SOURCES IN A TYPE B(U) PACKAGE

IV–51.   A ship carrying cargo, including a consignment of radioactive material, collides with a submerged object and sinks in territorial waters on a major shipping route to a depth of 30 m.

IV–52.   The consignment consists of a Type B(U) package in a freight container. The package contains three special form radioactive material ^{137}Cs sealed sources, with a total activity of 110 TBq. There are no other dangerous goods on board the vessel.

IV–53. While taking on water, the vessel notifies the appropriate notification point and the shipping company headquarters. All of the crew are rescued by a nearby vessel.

IV–54. The shipping company contacts the consignor of the Type B(U) package. The notification point and the consignor inform the emergency notification point of the potential emergency. The emergency notification point contacts the radiological assessors, the consignor and the carrier to assess the potential damage to the consignment and any possible radiological consequences.

IV–55. The consignor advises the authorities that containment of the radioactive material is ensured by the special form radioactive material capsules and by the Type B(U) package. While there is no suspicion of a release of radioactivity at the time of sinking, the radiological assessors estimate that corrosion of the capsules by sea water could lead to release of ^{137}Cs after a few months.

IV–56. Water samples are collected near the sunken vessel and show no contamination.

IV–57. The authorities and the consignor discuss the possibility of salvage and assess the time needed for maritime salvage operations on the sunken ship and cargo with a marine salvage company.

IV–58. Considering all the factors, including the popularity of tourism and fisheries in the area, a decision is made to try to salvage the ship and recover the freight container carrying the radioactive material within four months to limit the potential for corrosion. Regular monitoring and sampling of the marine environment in the immediate area is scheduled and conducted.

IV–59. The consignor and the public safety authorities work with the marine salvage company and are able to locate the freight container with the consignment of radioactive material. The ship is salvaged with the freight container three months after the sinking and moved to a nearby port. Marine environmental monitoring does not show any indication of contamination.

IV–60. After isolating the freight container, the general condition of the Type B(U) package is assessed and the package is surveyed for contamination. It is concluded that it is safe to move the package to the consignee's site, which is nearer than the consignor's, under special arrangement approved by the competent authority. The package is then shipped by road to the consignee's site.

IV–61.   Sampling of sea water and marine life is immediately performed in the area of the sinking, and six months later the absence of contamination is confirmed.

Potential consequences

IV–62.   An assessment of the potential radiological consequences via seafood consumption is performed by the radiological assessors to determine whether restrictions on food consumption are needed.

CONTRIBUTORS TO DRAFTING AND REVIEW

Aprilliani, D.	Nuclear Energy Regulatory Agency, Indonesia
Bajwa, C.	International Atomic Energy Agency
Breitinger, M.	International Atomic Energy Agency
Dodeman, J.F.	Nuclear Safety Authority, France
Garcia Alves, J.	Instituto Superior Técnico, Portugal
Hirose, M.	Nuclear Regulation Authority, Japan
Ito, D.	World Nuclear Transport Institute
Konnai, A.	National Maritime Research Institute, Japan
Marcotte, L.	Transport Canada, Canada
Mayor, A.	Office for Nuclear Regulation, United Kingdom
McBride, D.	Department of Energy, United States of America
Nandakumar, A.	Consultant, India
Presta, A.	World Nuclear Transport Institute
Sert, G.	Consultant, France
Tennant, R.	Canadian Nuclear Safety Commission, Canada

 **IAEA**
International Atomic Energy Agency

ORDERING LOCALLY

IAEA priced publications may be purchased from the sources listed below or from major local booksellers.

Orders for unpriced publications should be made directly to the IAEA. The contact details are given at the end of this list.

NORTH AMERICA

Bernan / Rowman & Littlefield
15250 NBN Way, Blue Ridge Summit, PA 17214, USA
Telephone: +1 800 462 6420 • Fax: +1 800 338 4550
Email: orders@rowman.com • Web site: www.rowman.com/bernan

REST OF WORLD

Please contact your preferred local supplier, or our lead distributor:

Eurospan Group
Gray's Inn House
127 Clerkenwell Road
London EC1R 5DB
United Kingdom

Trade orders and enquiries:
Telephone: +44 (0)176 760 4972 • Fax: +44 (0)176 760 1640
Email: eurospan@turpin-distribution.com

Individual orders:
www.eurospanbookstore.com/iaea

For further information:
Telephone: +44 (0)207 240 0856 • Fax: +44 (0)207 379 0609
Email: info@eurospangroup.com • Web site: www.eurospangroup.com

Orders for both priced and unpriced publications may be addressed directly to:

Marketing and Sales Unit
International Atomic Energy Agency
Vienna International Centre, PO Box 100, 1400 Vienna, Austria
Telephone: +43 1 2600 22529 or 22530 • Fax: +43 1 26007 22529
Email: sales.publications@iaea.org • Web site: www.iaea.org/publications